Clays and Health

Michel Rautureau
Celso de Sousa Figueiredo Gomes
Nicole Liewig
Mehrnaz Katouzian-Safadi

Clays and Health

Properties and Therapeutic Uses

Springer

Michel Rautureau
Docteur d'état ès sciences physiques
Maître de conférences des universités, retraité
Orléans, France

Nicole Liewig
Docteur d'État ès sciences naturelles
Chargée de recherche au CNRS
Strasbourg, France

Celso de Sousa Figueiredo Gomes
Ph.D in Materials Science
Professeur, université d'Aveiro (Portugal)
Aveiro, Portugal

Mehrnaz Katouzian-Safadi
Docteur d'état ès sciences biologiques
Centre national de la recherche scientifique
Chargée de cours, université Paris 7
Paris, France

ISBN 978-3-319-42883-3 ISBN 978-3-319-42884-0 (eBook)
DOI 10.1007/978-3-319-42884-0

Library of Congress Control Number: 2017935406

Printed on acid-free paper

This Springer imprint is published by Springer Nature
The registered company is Springer International Publishing AG
The registered company address is: Gewerbestrasse 11, 6330 Cham, Switzerland

Foreword

On reading this book a reader will certainly be astonished by the great age and importance of the observations on the biological role of clays. This empirically established knowledge, as that reported to us thanks to an expert author of Arabic medicine of the Middle Ages, resulted in the development of a simple and inexpensive pharmacopeia. Facing the poverty of a great part of the world's population and the crucial problem of hunger it is thus not surprising to see children eating clay. This "nutritional geophagia" certainy remained in many memories following the images broadcast in the media in the whole world showing Haitian populations making clay cookies. The astonishing multiplicity of clays will enable the readers of this book to make an enthralling comparative study of all their properties. However, the biological role of clays has only barely been studied by the scientific community. One of the primary reasons for this disinterest is probably the weak financial reporting regarding this pharmacopeia. Another explanation can be found in the hegemony of genome research that has characterized the study of living organisms during these last decades. It is likely that the awakening from a too-reductionist vision of biology and the development of a more integrative research will support the more functional approach required to study the biological roles of clays. But research teams will obtain the necessary human and financial resources only if the mindsets of those who allocate the resources for research evolve, in particular by giving higher consideration to those medicines known as "natural".

In this context, one of the great merits of this book is to sort out between the many biological qualities of the clays that have been established empirically since the beginnings of antiquity, even though the mechanisms of their actions are most often not perfectly understood, and those which for the moment do not appear to have any real basis and correspond to mere charlatanism. A better comprehension of the various biological roles of clays and their still poorly understood underlying mechanisms would then not only be of great interest for medicine and basic research, but it would also allow, vis-à-vis the explosion of false drug markets in poor countries and the booming of parallel medicine in the developed and secured countries, to better distinguish the contours and benefits of a pharmacopeia once again easily accessible.

One of the other merits of this book is to help the reader discover that geophagia is not only a human prerogative. Animals also learned how to benefit from clays that they select as for example, some Amazonian parrots that consume certain clays by taking advantage of their accessibility on eroded riverbanks. It is not common in fact that one can enrich the understanding of a pharmacopoeia developed by man from that which was also developed by animals in their natural environments during the evolution. Thus, whereas the human populations particularly developed the use of clay on a nutritional plane, considering their food benefit or their effectiveness in the case of intestinal disorders, clay ingestion by certain animals enabled them to feed on seeds that would otherwise be toxic.

In the face of sociomedical emergency, the proficiencies gathered in this work made it possible to cover the subject in its various forms of topicality and scientific utility and to provide the reader a synthetic, historical, and multidisciplinary vision on the relationship between man and animals with an omnipresent material on the surface of the planet. Although these reports are somewhat dated, readers will then be in the position of perceiving the modernity emerging from the instruments that enable today a scientific nanoscale approach to the cellular and crystalline structures, this scale where are precisely played the subtle exchanges and interactions between atoms at the border of life and mineral. The lesson to be drawn offers immense possibilities.

Membre de l'Académie des Sciences, Paris, France Yvon Le Maho
Ecophysiologiste, Directeur de Recherche au CNRS, Strasbourg, France
Institut pluridisciplinaire Hubert Curien, Strasbourg, France

Citations

Then the Lord God formed a man from the dust of the ground and breathed into his nostrils the breath of life, and the man became a living being.

Genesis, Chapter 2, Verse 7

I have always thought we were made of clay that life ever keeps on moulding us. Therefore the most important is to make sure that it does not become dry. Otherwise, all is lost: we become conceited, we stick stubbornly to our positions, we nestle down comfortably in our daily routine, and we build a barricade for ourselves forbidding us forever the possibility of being filled with wonder and enchantment.

J'ai toujours pensé que nous sommes faits de glaise que la vie n'a de cesse de nous modeler. Le plus important reste donc de faire très attention à ce qu'elle ne sèche pas. Sinon, tout est perdu: on devient prétentieux, on s'enferme dans ses certitudes, on se love dans son confortable train-train quotidien, on se construit un rempart nous interdisant pour toujours l'accès à l'émerveillement, à l'enchantement.

Maud Fontenoy, in *Le sel de la vie*, 2007

Preface

Humans have used clay throughout time. As a rule this material is always found close to water, and it can reciprocally indicate the nearby presence of water. In fact, in the history of our planet, clay has largely preceded humans who have taken advantage of its numerous and useful properties. If we consider that life has appeared and progressively developed in wet areas, while clay was being formed through destructive water action, it is fair to assume that over the planet's history a privileged relationship developed between living organisms and the very singular clay environment. This relationship supports the affinity of numerous animals for clays, including humans. Humans have progressively integrated clay in their culture and worship.

Clay, a popular material worldwide, has remained up to recent times scientifically poorly understood because it is constituted of minute particles (their very small particle size assigns clays into the field of colloidal particle properties; clay can be found as slurry, paste, and mineral dust, that is, in states that have long been unfriendly to the allocation of specific traits). Following the first scientific studies it has become pretty clear that clay is constituted of elementary particles whose dimensions, as a rule, are less than 2 μm (2 $\mu m = 2 \times 10^{-3}$ mm). This size criterion is actually a frequently mentioned fundamental factor although it is not exclusive to clay inasmuch as other minerals can display similar sizes. Contrary to most common minerals, clay particles rarely display regular shapes visible to the naked eye, nor do they show a permanent solid state. Despite these peculiarities, a priori unfavorable to its applications, clay can easily be recognized and its use generalized, whereas the properties of numerous other minerals that could be easily manipulated were only recognized much later. Two recently published books on this point present an excellent synthesis of the general medical applications of minerals: *Minerals and Human Health: Benefits and Risks/Os Minerais e a Saúde Humana*: *Benefícios e Riscos* by C. Gomes and J. B. Silva (2006), bilingual edition of the authors, Multiponto, Porto; and *Mineralogia Aplicada: Salud y Medio Ambiente* by M. I. Carretero and M. Pozo (2007), Thomson (Editor), Madrid. Two recently published articles deserve to be studied as well: "Clay Minerals and Their Beneficial Effects Upon Human Health: A Review" by M. I. Carretero (2002), *Applied Clay Science*,

21(3–4):155–163; and "Minerals and Clay Minerals in Medical Geology" by C. Gomes and J. Silva (2007), *Applied Clay Science*, 36:4–21.

The existence of clay in geological sedimentary areas provides this material with a particular role relative to water; clay is constituted of hydrous phyllosilicate minerals that are mostly derived from the weathering action of water on geological materials occurring at or near the Earth's surface. Therefore, there are distinctive species of "clay minerals" in nature. Thus in this book we use either the singular to designate any individual or species within the overall existing clay minerals, or the plural to mean either a certain group of clay minerals or the clay minerals in general.

The origin of clays is essentially related to the degradation of diverse pre-existing rocks. Clays keep visible traces of their origin inside their crystalline structure and the physicochemistry in their particles: their size is always small, and their chemical composition and the atomic organization are always determined by the parent material. Clays are unstable and sensitive to any new environment where they coexist, although this does not grant them a definitive stable state. They react to very weak changes of energy and constitute a privileged intermediate state with an infinite number of natural reactions. Thanks to the extraordinary progress of scientific methods, the studies being carried out show the existence in these minerals of numerous defects (mainly formed by substitution of atoms and limitations of particle growth) relative to the ideal or theoretical model. Their physical and chemical properties and specificities are evidence of the initial heritage acquired in their original environment or of later adaptations connected to the eventual evolution of this environment over the course of time.

> *Clays are the essential components of the "original soup" frequently cited in scientific papers, from which every living organism has been formed. If clays have not truly been at the origin of life, we can assume that they have constantly been close to and accompanied it. They have contributed to make life possible and been effective performers and efficient partners. Civilizations and their religions have not been misled; all of them make spontaneous references to clay and to water, with which clay is always closely related.*

Now, we can consider clays as a family of minerals which are very sensitive to environmental conditions. They can be considered as "structures in constant evolution". Consequently, the fundamental research on clays is particularly dynamic and the prospects for their applications offer an infinite variety. Their specificities are very favourable to applications in the health field, where health and tradition finally could cross. In order to describe clays we should remember two important definitions:

1. The first is granulometric: They consist of particles whose sizes are, as a rule, less than 2 μm.
2. The second is mineralogical: They are hydrous phyllosilicates.[1]

In the domain or field of human health (and of animal and plant health as well), the subject we deal with in this book, we should ask ourselves about the interaction between clays and the numerous living organisms of all types. Knowledge of the interaction mechanisms between living organisms and clays remained empirical

[1] Phyllitic formed by a morpheme phullon, from the Greek *φυλλον* which means "layer shaped".

until the beginning of the twentieth century; before then no solid scientific study was available or even possible. Physicians have progressively acquired and transmitted some very important practical information, although limited by the lack of an effective analytical methodology. Chemistry and physics have progressively provided the essential quantitative measurements confirming the crystalline mineral structure of clays, that is, the organization of atoms in a regular network. Since then, research has progressed rapidly, first in the industrial field, and lately in the less easily studied biological interface fields. Nevertheless, this scientific progress has not been accessible to everyone interested in the use of clays. Numerous interventions still remain totally empirical and undemonstrated or explained, a situation that could sometimes be dangerous. However, today there are sufficient scientific fundamentals to allow us to explain some positive effects of the use of clays in health. Sometimes the explanation is uncertain or even absent, but we have to admit the reality of the unanimously recognized positive effects, and postpone understanding them and their scientific explanation until later. The opinion of physicians evolves as shown in a recent book entitled *ABC de l'argile* by J. C. Charrié (2007), Grancher, Paris. Dr. Charrié is a general practitioner who tries, along with some colleagues, to find explanations of the traditional use of clay in scientific papers in order to justify its clinical use based on the integration of traditional information in science.

For the present it is important to express one reality firmly: with regard to living organisms, in general, clays are not healing or "curative," and we should not only remember the benefits. Even if some clays has useful properties, in some precise cases they can be hazardous too, such as chrysotile (a phyllosilicate that is part of the composition of asbestos, a technically very useful industrial mineral whose carcinogenic action on the pleura of the lungs is related to the fibrous shape of its particles). The term "asbestos" means a finished industrial and commercial product consisting of different acicular minerals (in the form of very fine fibers). These minerals have a special fibrous morphology and are resistant to fire; they are often called asbestos by mineralogists. Chrysotile (or white asbestos) [$Mg_3\,Si_2\,O_5\,(OH)_4$] is a frequently used mineral, and is also classified as a serpentine phyllosilicate. One uses also amosite (brown asbestos) [$(Fe, Mg)_7\,Si_8\,O_{22}\,(OH)_2$] and crocidolite (blue asbestos) [$Na_2\,Fe_5\,Si_8\,O_{22}\,(OH)_2$] which are not phyllosilicates; these minerals belong to the amphibole family.

There are several ways to present clay properties; we use two fundamentally distinctive methods based upon the concepts of order and disorder, two fundamental stable states of matter. Because clay particles have a microcrystalline character, researchers have rapidly directed their arguments by first taking advantage of the crystalline state characterized by order; and lately by attributing a more important role to the microparticle itself. They have tried to understand the properties of certain assemblages, comprising a large number of clay particles: it is the disorder aspect that has been taken into account. Clay is formed of microscopic crystallites therefore the mathematic modeling methods could be applied to these two physical states of order–disorder, but that does not mean either an easy undertaking or that the obtained results could be easily extended from the microscopic to the macroscopic scale.

Oddly enough, clays simultaneously respond to two criteria related to very distinctive dimensional domains. Order exists at the microscopic scale; apparent disorder is seen at the macroscopic scale. Therefore, the definition of the representation scale is fundamental, inasmuch as there is a border zone, one kind of dimensional fracture, enabling each side to show different properties, and thus it is very difficult to find one justifying connection between these two states. It is constantly important to present applications that can be expressed and explained in both domains, sometimes through specific properties of both domains. This duality of order–disorder leads to the distinction of two main types of mechanisms, which direct the therapeutic applications of clays.

The order and disorder represent the true reality of natural states fairly well: between both matter states adaptability allows considering intermediate unstable states, frequently reversible and unstable. Nature needs stabilizing elements in order to maintain its stability dynamically over time, and for its renewal it needs sufficiently flexible and adaptable systems. Clay minerals show these two qualities. Their well-organized atomic structure is able to resist attacks taking place over time, sometimes over millions or even billions of years. Nevertheless, their great affinity for water and the presence of inherited defects related to their genesis allow their response, and rapid adaptation to solicitations that would have no effect on other compounds. Therefore, most clay properties depend upon their particular physicochemical properties, which favor reactions of assemblage, of absorption and adsorption, and of exchange. Their high absorption capacity allows the fixation of numerous substances (cations or molecules), some of them toxic, a fact that gives them therapeutic capacities, particularly in the gastrointestinal system. Most of these mechanisms only require weak energies, with a frequently reversible intermediate state, similar to a catalyst reused almost to the infinite. A fundamental point to understand the mechanisms of clay minerals' action is their ionic constitution and their association with the compensating exchange of ions, or to the mobile molecules frequently involved in the main exchange reactions required and utilized in therapeutics.

Whenever clays are in contact with biological material (possibly alive), they represent two worlds and two completely different types of existence that encounter each other. This results in a great diversity of properties and applications. In order to understand how clay works it is important to consider that clay is always ready to modify and evolve itself to adapt and find the best balance.

Organisms or living beings are sensitive and reactive, some of them conscious. Our aim should not neglect the sensorial and psychosensorial aspects of the contact between a conscious living body and clay; in addition, higher living beings such as humans have the possibility of adjusting their responses to a certain perception. The expression *sweet clay* acquires a real dimension here: "We react well if we are well." A kind of symbiosis between clay and the body takes place. Associated with water, and at an appropriate temperature, clay possesses this complementary property of setting in motion the perception of well-being.

Numerous and diverse traditional uses of clays have been known for at least 10,000 years. All civilizations with access to clays and having applied them were deeply influenced. From the beginning of the twentieth century onwards we began to have a better knowledge of the properties of these minerals at both the micro- and macroscopic scales. Within the domain of their therapeutic use we can say that the scientific advances will bring about useful developments through the linkage of one classic chemical or physicochemical associated (or not) with a complementary and efficient psychotherapeutic effect.

At a time when science and technology produce marvellous useful materials, clay still continues to have remarkable efficiency in spite of the longevity of its uses. Presently its applications range from civil engineering works up to the nanotechnologies. In fact it is at this small scale that one can find the key to health applications which are dealt with in this book.

We limit our ambitions to recognized facts having properties that can be identified and quantified. Of course, the absence of scientific knowledge has led certain authors to establish cause-and-effect relationships without any solid significance. However, this does not mean, a priori, that their observations are without justification: it is the attribution being put forward that deserves to be more precise. We have chosen four examples from the literature, but there are many others:

1. *"Radioactive" properties of clays:* Phyllosilicates by themselves do not have such properties. However, clays mainly have their origin on the alteration of silicate rocks and thus have to obey the rules of erosion and detritism. Consequently sometimes they can be mixed with uranium compounds derived from the same rocks, which possess radioactive properties. Because such compounds are finely divided in terms of particle size, they could be separated with difficulty from clays, in all cases not naturally. In this way clay commercial products are weakly radioactive (we emphasize that this property is rarely measured and indicated on the clay packages.)
2. *"Magnetic" properties:* Those currently cited are not identified in clays. In fact, they result from the presence of fine particles of magnetic metal compounds that are mixed with clays.
3. *"Antibiotic or antiseptic" properties:* Although clays are frequently cited as possessing them, they do not have these properties, but because of their sheet-based structures they can constitute impermeable barriers to water and air in particular; it is obvious that they can preserve the organism from contact with a polluted environment and keep oxygen from reaching the surface of an infected zone thus prohibiting the development of certain microorganisms. Another property allows them, by adsorption, to assess adsorption molecules that are really "antibiotic," but are only associated with clay. The confusion in these cases is always the result of ignorance of the complex mechanisms involved, which should be analyzed case by case, without generalization.
4. *Clay color:* This is frequently a commercial argument that is considered to be important; it is not systematically linked up with known properties, and is never the exclusive cause.

It should be noted that when radioactivity in clay treatment was first spoken about, the word "radioactivity" did not have the meaning it has today and there has been a semantic shift. The same applies to the word "magnetism".

The authors' goal is not to fight these forms of obscurantism, but to contribute to a better interpretation of reality in order to make better use of clays thanks to understanding the way they build up and the way they work. The difficulties come from the microscopic character of the implied mineral phases, knowing that scientific studies allow a progressive correct interpretation of their properties. In this evolution clinical observation cannot be disregarded; nowadays the latter is frequently forgotten, but through the impacts it has upon life, it allows the orientation of analytical research and opens a way to new scientific perspectives.

To sum up, the authors' goal is not turned towards a controversy aimed at what cannot be verified but is focused towards well-established scientific knowledge that is likely to evolve, and resolutely turned towards an enlarged use of these particular minerals. Their application in new methods of biomineralogy is presently being developed by research laboratories. Clays have almost always been characterized by the concept of balance. This notion of balance is also a harmonious factor in life: the quest for balance allows natural activities to occur and act so that all the vital functions of an organism participate in its life.

In the health field, the use of clay is too ancient to be completely supplanted by supposed modern scientific or medical pharmaceutical methods.

Clays are components of the mineral kingdom and humans have almost always been living in contact with them. Water was the main actor in the formation of clays during the Earth's geological formation process; water remains the main source allowing life to exist. Life appeared in contact with these two elements. Therefore the strong interaction between clays, water, and living organisms is not surprising. Among the living organisms human beings have the advantage of being able to arrive gradually at an understanding of the interaction mechanisms between clay and humans.

If scientific medicine did not exist, empirical medicine would be the only one at our disposal, as it has been for a long time. Under certain conditions, it is still being used and will probably continue to persevere. Mankind has the chance of possessing these two very different approaches of medicine that can be mutually enriching.

Long before being able to explain the various properties of clays and their interactions with health, humans very quickly recognized their utility. In the absence of scientifically structured thought, they initially transposed on a mystical level what they could observe. Legends and myths were thus born, often rather close to reality. Humans, conscious of their existence and their mortality, always felt the need to know their origin.

The progressive evolution from legend to a desire for knowledge occurred naturally, bringing with it a religious approach. Thus, almost all religions contain traces of this heritage. If we can clarify this approach today, we must recognize that the existence of these beliefs is little by little getting closer to our scientific model. This evolution does not remove anything from the spiritual interest but makes it possible to give it a status closer to reality. Thus, the legendary world enters into the unfolding of history.

"Man is made of clay." That is a vista opening a vast cultural, religious, or simply vital domain of reflections.

This development is found historically in the use of clay by humans. Today we can see a transition towards a scientific expression of these uses. For many people it is, however, quite difficult to forget this apparently magic aspect of clay, symbolically named "ground" or "earth." We do not try to reverse this tendency but simply hope to offer all those who express the need or the wish an access road to the scientific methods. Let us not forget that anything which allows a real return to the balance and harmony of health cannot be neglected: at least the many properties of clays permit this choice.

Orléans, France — Michel Rautureau
Aveiro, Portugal — Celso de Sousa Figueiredo Gomes
Strasbourg, France — Nicole Liewig
Paris, France — Mehrnaz Katouzian-Safadi

Abstract

This monograph aims to disclose the importance of clay and clay minerals for human health and well-being. Such importance is based upon the textural, physical, and chemical specificities of clay and clay minerals that justify their use by humans over time, either in the form of external or topical applications, or in the form of internal applications. Naturally these applications were first simply based on empiric experience, and only much later on both scientific and technical experience, particularly thorough crystallochemical and medical knowledge.

Clays and clay minerals (their fundamental constituents) widely occur at the Earth's surface, making up an essential part of the soils that are of paramount importance for life in general, and for human life in particular. There is a close relationship between the human and clay habitats, a close relationship extensive to clay and human body properties and to their interactions with water and air. Both clay and the human body are chemical systems constituted of the same chemical elements, although in different proportions. In the field of health clay can be utilized either in the form of dry powders, or in the form of clay–water dispersion or suspension, or in the form of clay–water paste. (Mud therapy and peloid therapy take advantage of the thermal, chemical, and rheological properties of the clay–water system, the water being sea or salt lake water or mineromedicinal water.)

Therapeutic properties result from various modes of reactivity between clay minerals and the human body. Clay therapeutic properties are particularly based on the ability of clay minerals to adsorb and retain harmful and toxic substances that justify its use in the treatment of gastrointestinal disorders, the deliberate eating of clay or clayey soil (geophagy) being practiced worldwide since antiquity. Clay can also act as a protective agent.

It is well known that clay can function as active principles or excipients in pharmaceutical formulations. There is also well scientifically supported information regarding the antibacterial action of certain clays. However, it is now known that certain interactions and reactivity could be deleterious to human health; for instance, the inhalation of fibrous clay minerals can cause irreversible harmful effects of the pleura.

Clay's positive and negative effects have given rise to many legends and historical records and many reports can be found as well in the more or less recent specialized scientific literature.

Ingestion, inhalation, and dermal absorption are the three pathways of clay–human body interaction.

Cosmetics is another interesting field of study for clay, taking advantage of clay's positive interaction with human skin due to specific textural, mineralogical, chemical, and rheological properties; also, the use of clay in cosmeticeuticals represents an important market.

These days the theoretical and practical studies of both types of reactivity referred to are dealt with in an emergent scientific field called medical geology which is characterized by high interdisciplinarity and is progressively becoming more and better established. Recent developments and advances concerning the interaction of clay and clay minerals with human health open and meet interesting challenges by mineralogists, geochemists, toxicologists, and biomedical and public health scientists among others.

The monograph is organized in nine chapters, as well as several appendices, whose topics and titles are referred to: background; definition of clay and clay minerals; historical data of a natural pharmacopeia: clay in the corpus of "the medieval pharmacopeia" written in Arabic; geological and mineralogical information and description of clay and clay minerals; clay reactivity depending on the crystallochemical properties of clay minerals; general information on clay applications in the fields of health and well-being; therapies based on clays; principal modes of clay use; and provisioning, recycling, and trade of clays.

Contents

Acknowledgments

The authors thank, for their participation or their counsel:

Ms. Jade Allègre
Ms. Nadia Alem
Mr. Alain Baronnet
Mr. Daniel Beaufort
Mr. Dalil Boubakeur
Dr. Jean-Christophe Charrié
Mr. Norbert Clauer
Ms. Joelle Duplay
Ms. Francoise Elsass
Ms. Josiane Francart
Mr. Hervé Gaboriau
Ms. Khadidja Guellil
Ms. Anne-Lise Haenni
Ms. Catherine Rodger
Mr. Toshiro Kogure
Mr. Mohamed Khodja
Ms. Anne-Marie Jaunet
Mr. André Lingois
Mr. Alain Piquard
Mr. Guy Pierre
Mr. Alain Plançon
Mr. Roshdi Rashed
Ms. Thérèse Rautureau
Ms. Joëlle Ricordel
Mr. Eduardo Ruiz-Hitzky
Mr. Henri Van Damme
Mr. Graham and Dorothy Caïrn-Smith
Mr. Fernando Veniale

Mr. Jacques Yvon
Mr. Jorge Hamilton Cardoso Gomes
Mr. João Baptista Pereira Silva
Scientific societies: GFA and AIPEA
Mr. Claude Gérard for the revision of the English presentation
This book is a tribute to Simonne Caillère, Stéphane Hénin, and Georges Millot.

Chapter 1
Background

Abstract There is a long history of using clay to treat human health, physical balance, and psychophysiological balance. The literature is full of empirical examples of clay uses, supported by traditional methods often perpetuated in the form of legends related to recovery. Numerous cultural or worship practices have supported these uses, with or without solid justification, but always for reasons close to vital concerns. In popular tradition it is the *earth* (in English), the *terre* (in French), that has a powerful part in supporting life. It is the task of modern scientists to distinguish among the many uses of clays, those that are truly explainable and those that are less logical or even doubtful. It is the job of scientists to integrate tradition into science.

For ages the principal uses of clays were related to the digestive function and its most current disorders and also to dermatology. To fulfill these practices, significant amounts of clays are now necessary; they must also contain homogeneous and consistent properties. The generic term of clay includes many families or typologies based on their mineralogical description. The best characterized and most common clays are explored in unremoulded and unpolluted important deposits. Historically the most commonly used clays have characteristics that depend on the major clay minerals present in them: kaolinites (often white); smectites (often green); illites and talcs (often white); and palygorskites and sepiolites, both characterized by the fibrous shape of their particles (often white or nearly white). Because clay deposits are primarily sedimentary, various types of clays are often present simultaneously in such deposits. In almost all cases, nonclay minerals are also present in small quantities in the form of very small particles associated with clay minerals. These trace minerals can also have an important but often ignored role, such as in the case of amorphous silica.

Thus we develop the essential basic concepts concerning the morphology and the crystalline structure of these various clay minerals that are more scientifically called *hydrous phyllosilicates* which is more correct from the mineralogical point of view. The designation means "sheet silicate or layered silicate minerals characterized by possessing structures made of layers of $(SiO_4)^{4-}$ tetrahedral units linked together to form flat sheets with the composition $(Si_4O_{10})^n$, and that contain water molecules or hydroxyls."

The classification of phyllosilicates, clay minerals included, is discussed here in order to establish their active characteristics clearly. A short chapter is devoted to the various types of structures and to the methods of extraction, preparation, and purification of the clays, without forgetting their recycling. The associated natural

M. Rautureau et al., *Clays and Health*, DOI 10.1007/978-3-319-42884-0_1

minerals are dealt with as well as the more complex preparations involving natural clay such as the traditional "peloids."

Peloid comes from the Greek σοληπ or *pêlos*, and in general indicates a therapeutic mud made up of a mixture of mineral water or seawater associated with mineral or organic matter. A clay paste can thus be included in the definition of a peloid when it is used for therapeutic purposes. The term "peloid" is employed by the spa therapy industry or thermal resort medicine and evokes, for example, the peloid of Dax or the peloid of Rochefort indicating the origin of the clayey sediments. Another word characterizes this aspect: *pêlodès*, which means muddy.

Therapeutic properties result from various modes of reactivity between clay and clay minerals and the organism. They vary with the morphology and the crystalline structures specific to the various families of the existing clay minerals, as well as with their modes of preparation and application. Clay can be modified, even destroyed, when it comes in contact with various environments. For example, the acid pH of the stomachic medium can promote an acid attack of clay, leading to a residue primarily formed from the ions located at the center of both octahedrons and tetrahedrons (especially of silica in various forms, and oxides or hydroxides of aluminum). A great number of geochemical studies were undertaken to follow the evolution of clays in natural environments. We can take advantage of these study results to compare these evolutions with those of the same minerals placed in contact with a living organism.

When clay is not destroyed under the effect of stomach acids, it can constitute an excellent protector that can act by absorption of toxic molecules in the vicinity of mucous membranes and the skin. We can cite the action of poultices made up of a very great number of clay particles and associated elements. Thus we have to distinguish in more detail the concepts of adsorption, absorption, and exchange. Moreover, the swelling property of certain clays is also of great importance in how they are used; it constitutes a fundamental parameter to control during the internal uses of clays.

The relation between clay and water (the liquid that is the main component of the human body) is fundamental to understanding and explaining the variable properties of the clay–water mixture. Clay can indeed react permanently with water to form more or less hydrated units (dry powder, paste, mud, suspension) to which specific properties and applications correspond. In addition, a pasty clay–water mixture is relatively impermeable, which imparts very particular and important properties to it.

The great number of parameters to be controlled, not only on the level of clay itself but also on the level of the individual who uses it as personal medicine, is at the origin of a multitude of applications. These methods have in common the crucial step of the initial preparation of clay, followed or not, of what is now called "maturation" conditioning. This term is analyzed in detail taking into account the importance of this stage in the success of the therapeutic use of clay. It can be found included in mixtures, for example, with essential oils. The principal operating process is the diffusion of molecules in the medium (air, water, oil, etc.) that contains clay. The development of microorganisms can play a determining role.

The action of clays is often associated with a barrier effect. This effect is relatively complex and comprises direct and indirect aspects. In certain cases, the clays applied on more or less weakened skin constitute an effective "mechanical barrier" impermeable to air and water, able to isolate the surface from this weakened organ that, without this protection, would be put in direct contact with harmful compounds. Sometimes the simple protection relative to the access of germs is itself beneficial. In other cases, such as in the field of bacteriological contaminations, clays can act as a barrier to air and thus make it possible to fight efficiently against the development of aerobic bacteria (which require oxygen). Furthermore, clays can be regarded as chemical reservoirs of elements or molecules that enable them to take part in chemical reactions and exchanges while continuing to play the part of a physical barrier isolating the zone being treated from the external environment. Used in cataplasms they can allow exchanges of ions through the tissues. Important disorders such as deep ulcers or Buruli ulcers can thus be effectively treated by clays. In applications such as stomach disorders, clays at the same time make it possible to isolate an organ (intestinal mucous membrane) from an inappropriate gastric medium and simultaneously to eliminate toxins.

In the field of health protection, clays are also indirectly and successfully used to filter water polluted by bacteria. In this same health field, focused on food concerns, it is known that French soldiers in the First World War who had received mustard (a condiment containing a small quantity of clay as a jellifying agent) in their food ration were less subject to the effects of dysentery and other prevalent diseases within the armies in the field. Also the military kits of Russian soldiers automatically included mustard. Such examples prove the effectiveness of clay's action.

Considering the diversity of clay uses, whatever the pathology, it is necessary to consult a doctor in order to decide on the most adequate therapeutic procedure. This medical diagnosis (or that of a health care professional) is also desirable for aesthetic care or for well-being; thermal treatments clearly belong to the medical field.

The use of clays in skin and beauty care is very old. There they have a crucial role as mineral filler, active or interactive. They have a very soft abrasive effect. Their properties of adsorption and ion exchange make it possible to associate many active elements with them such as, dyes, perfumes, essential oils, and trace elements, among others. The term "trace elements" is often employed; it indicates chemical elements in very small quantity but necessary to a vital activity.

The stability of clays allows their use in mixtures (creams, gels) to which they confer thixotropic and rheological properties that characterize the behavior of a plastic paste (rheology) and the jellification of a suspension (thixotropy). Sometimes it is necessary to modify the properties of natural clay through the incorporation into the interlayer space of ions or polar molecules that allow modification of the medium's viscosity. For example, chains containing a functional ammonium (alkyl ammonium type) can be used for this purpose.

Clay can also be used as a vector to transport ions or molecules to their target in (or on) the organism. Many ointments, creams, and shampoos benefit from this property. The organization of the clay particles in an ointment brings some energy into play. When deposited on the skin, it generates surface stresses that have a determining role and can be used to regenerate the skin (for instance, wrinkles and dryness).

When clay is employed in significant amounts, reusing the same material is sometimes possible for the same person; certain rules of hygiene must be followed, in particular taking into account existing laws. This method should not be encouraged. After clay use, and passage through an autoclave, recycling is possible by reintroducing the clay mineral matrix into the environment without biological pollution risk. In the case of need or of doubt (viral or bacteriological, e.g.), treatment in an autoclave or exposure to ultraviolet rays is able to neutralize the majority of the microorganisms and, if necessary, clay calcination (between 600 and 1200°C) definitively destroys all the organic matter by producing a ceramic composed of high-temperature minerals. There is always a possible solution to recycling, or neutralizing, clay after use.

The action of clays is generally associated with the nature of the liquid used to disperse them: mainly water but other liquids can also ensure this dispersion. Dry clay is seldom used, except for powdery talc. To underline the importance of this concept, recall that in agriculture any ground constituted only of dry clay is practically uncultivable. Better still, a very clean mass of clay is practically sterile in microbiological terms. It is this property that is used to insulate and treat, often successfully, infected or superinfected surface wounds.

From an economic point of view the use of clays is an important activity: numerous uses of clay depend upon diversified domains. It is advisable to be very vigilant of the state of the geological reserves of this raw material: they require the existence and safeguarding of many uses. These microcrystallized minerals are renewable natural products on a geological timescale, but very slow compared to the human timescale. To guarantee provisioning continuity it is advisable to have a clay deposit of good quality able to accomplish the exploitation objectives (within the limits of deposit extension). The use of these minerals is barely controlled by legislation. However, in the event of specific danger being recognized, they can be prohibited, such as in the case of chrysotile (one of the components of asbestos as mentioned in the preface).

On a medical basis, an important quantity of clay is used in self-medication and in natural alternative medicine, but use by the general medical community is also important. These two aspects are complementary. Gradually the medical community, pharmacists, and the authorities are reconsidering clay as a possible active element as an alternative to the use of drugs produced by the chemical industry.

Chapter 2
Clay and Clay Mineral Definition

Abstract Any science, scientific field, or scientific domain requires a well-established and defined object of study as well as appropriate methodologies to develop knowledge about that object. However, over time the definition or concept of the object of study can undergo modifications and adaptations due to scientific advances, both in theoretical and experimental terms.

Clay science is multidisciplinary with the object of study both clay and clay minerals. It receives important contributions from other sciences or disciplines, such as mineralogy, crystallography, chemistry, geochemistry, sedimentology, geology, pedology, agronomy, soil mechanics, colloid chemistry, materials science, biology and biotechnology, medicine and public health, pharmacy, geoengineering, and environmental engineering. As in other sciences the object or objects of study referred to require a definition that would deserve general acceptance by all those interested in it, both theoretically and practically, despite the historical evolution of the concepts. Different from other sciences, the enormous diversity of the natural material denominated clay, in geological, compositional (physical, chemical, and physicochemical), and technological terms, as well as the enormous diversity of clay fields of interest and applications, may justify the lack of consensus on clay and clay mineral definition that is shown in this chapter.

2.1 Introduction

Through the ages clays have received definitions adapted to the state of knowledge at the time and the uses to which they were put. As universal materials they cannot be limited to only one definition. The vocabulary of every language is rich in characteristic and descriptive terms; all these minerals are common to all our activities. Near the word "clay," a generic universal name is that of "earth," associated with qualifying names such as potter's clay, fire clay, or green clay. The term "clay" comes from the Greek word *argilla*, the root of which, *argos*, signifies "white," the color of kaolin clay used in ceramics. The relation to purity was quickly established with this color even though most clays are not white.

Clay science is less than 100 years old, and during its evolution the definition of clay and clay minerals, the objects of study of this science, has shown significant changes that have been proposed, developed, and adopted by several authors and

M. Rautureau et al., *Clays and Health*, DOI 10.1007/978-3-319-42884-0_2

clay groups or clay societies. Such changes were founded not only on basic but selective criteria heavily dependent upon the particular academic education and professional experience of the authors, for instance, soil scientists, mineralogists, geologists, or material scientists, and on scientific advances, particularly provided by the increasing capability of the analytical instrumentation and methodologies, informatics, and computing.

Clay occurs where we less expect (Moll 2001), not only in rocks of any type (sedimentary, igneous, metamorphic), but also dispersed in atmospheric aerosols (Sanfeliu 2002) and suspended in the water of oceans, seas, lakes, lagoons, and rivers (Gomes 1988). Naturally and industrially produced dusts can contain clay minerals, that could have deleterious effects on human health.

There are several types of clay based on geological, technological, and commercial arguments: primary or residual clay, secondary or sedimentary clay; swelling clay, plastic clay, absorbent clay, refractory clay, nanoclay, commercial special clay (ball clay, bentonite, kaolin, Fuller's Earth, fibrous clay, fire clay, flint clay), and commercial common clay (pottery clay, heavy clay). *Special clays* relative to *common clays* are characterized, as a rule, by the smaller number and size of the deposits, simpler composition (although requiring more complex and expensive processing), and higher commercial value of the manufactured products.

Clay is also a major constituent of the so-called *healing clays* and *edible clays*. Within the healing clays some are used solely for external applications with therapeutic and/or cosmetic purposes, either in the natural environment, or inside specialized spas, as are the cases of *natural peloids* and *peloids s.s.* (*stricto sensu*). Other healing clays, the so-called edible clays, are used for internal or oral applications.

Each clay has its own *clay fraction* in which the clay particles and clay minerals—the essential constituents of clay—are more concentrated. Conventionally some disciplines set a maximum size for clay particles: pedology <2 μm; geology, sedimentology, and geoengineering <4 μm; and colloid science <1 μm. However, the upper limiting size of an individual particle or crystal for clay minerals has not been established.

When the optical microscope was the only instrumental tool to analyze geological materials, that is, during the last decades of the nineteenth century and the first decades of the twentieth, clay was defined as an association of fine-grained minerals with particle diameters of less than 2 μm, that is, beyond the resolution of the optical microscope.

Earlier chemical analysis became a fundamental tool to determine the chemical composition of both clays and clay minerals, allowing the establishment of their typology. Soil science and colloidal chemistry and the development of analytical techniques, such as differential and gravimetric thermal analysis, transmission and scanning electron microscopy have contributed a great deal to the identification and quantification, as well as the explanation of clay and clay minerals' basic properties.

Starting with the third decade of the twentieth century X-ray diffraction became a powerful tool for investigating the structure of fine-grained materials, clay minerals included, allowing the differentiation of clay mineral structures, an achievement fundamental to explaining their distinctive properties. Many other analytical methods that became available, such as scanning electron microscopy (SEM) associated with X-ray energy dispersive spectrometry (EDS), and synchrotron X-ray diffraction (SR-XRD), have played essential roles in modern clay research.

2.2 Definition, Classification, and Nomenclature of Clays and Clay Minerals

After the works of precursor researchers, both the AIPEA (Association Pour l'Étude des Argiles) and CMS (Clay Mineral Society) Nomenclature Committees have been involved for several years in the definition, classification, and terminology of clays, clay minerals, related materials, and specific properties. For instance, the AIPEA Nomenclature Committee has proposed several reports with recommendations: Brindley and Pedro (1970, 1972, 1976); Bailey (1980, 1982, 1989); Bailey et al. (1986); Guggenheim et al. (1997, 2002, 2006). Recommendations have also been proposed in joint reports of the AIPEA and CMS Nomenclature Committees: Guggenheim and Martin (1995, 1996), and by the CMS Nomenclature Committee: Brindley et al. (1968); Bailey et al. (1971a,b, 1979); Martin et al. (1991).

Despite the continued efforts of these nomenclature committees, the definitions of clay and clay minerals still are not unanimous. This can be understood because clay is most probably the mineral resource exhibiting the highest variability in geologic, textural, and mineralogical terms and the most diverse applications. Thus either the professionals or the disciplines interested in clay try to enhance the property or properties relevant for a particular application in its definition, situations well expressed by Gomes (1988, 2002):

> *1—Geologists* consider clay as a geologic product of generalized occurrence and of fine granularity that occurs at the surface or near the surface of the Earth, and that is formed at the interfaces between the earth crust and the atmosphere, hydrosphere, and biosphere as the result of the mechanical and/or chemical alteration of rocks; also *geologists* classify clays on the basis of their origin and occurrence: residual or primary and sedimentary or secondary; marine, alluvial, glacial, eolian, etc.; *2—Mineralogists* consider clay as an aggregate or mixture of minerals of fine granularity, consisting principally of clay minerals, which are hydrous phyllosilicates based on Si, O, OH, H_2O, and that elements such as Al, Mg, Fe, K, Ca and Mg can participate on their composition; *3—Petrologists* consider clay as a rock, as a rule weakly consolidated, formed by very fine mineral particles non identifiable at naked eye or using a magnifying glass; *4—Sedimentologists* consider clay as a granulometric term that identifies the sediment fraction composed of particles with e.s.d. (equivalent spherical diameter less than 2 μm); *5—Civil and geotechnical engineers* consider clay as the less than 4 μm fraction of the soil and classify clays or clayey soils as swelling or non swelling, soft or hard, that is selecting the most relevant properties that should be taken into account when the foundation of housing and public construction works is based on them; *6—Ceramists* consider clay as a natural material that whenever mixed with water in adequate quantity becomes plastic allowing its workability and shaping, and that hardens and maintains the acquired shape after drying or firing; ceramists classify clays

on the basis of their plasticity, firing properties and uses: low plasticity, medium plasticity, high plasticity; white-burning, red-burning, high or low refractory; common clay (brick clay and pottery clay); special clay (china clay, fire clay, ball clay, fibrous clay); *7—Pedologists* consider clay as the active fraction of the soil that comprises particles of colloidal size (<0.1 μm) that results from the action of pedological processes upon rocks, and which are responsible for the reversible fixation of cations and anions, such as NH_4^+, K^+, PO_4^{-3} and NO_3^-), etc.; *8—Agronomists* consider *clay* as being the colloidal fraction of the soil, important in terms of structure, texture and composition for vegetal fixation and growth; *9*—even the *non educated people* consider clay as a type of earth that gets slippery when wet, and that could be moulded or shaped.

With regard to clay minerals' definition, classification, and nomenclature, over time there has been some controversy as well.

Clay minerals occur in rocks of all lithologies and ages. Also, clay minerals are common and important constituents of the fine fraction of clay as well as of the fine fraction of soils that are fundamental for life in general, and for human life in particular.

Clay minerals are fine-grained hydrous phyllosilicates, minerals of planar or nonplanar structures that could impart plasticity (if the right amount of water is added) and hardening (if dried or fired) to the clay in which they occur. Clay may contain other materials that impart plasticity and harden when dried or fired. Clay minerals are not defined a priori as fine-grained minerals because clays are fine-grained, but rather they may be of any crystallite size so that the term "clay mineral" is consistent with the definition of mineral, which is unrelated to crystallite size. The particle-size issue has been separated from the clay mineral definition.

As a rule, in addition to clay minerals, clay may contain the so-called *associated minerals* (silicates such as quartz and feldspar; carbonates such as calcite, oxides, and hydrated oxides of iron and aluminum such as hematite, goethite, and gibbsite), X-ray amorphous materials such as organic matter, and inorganic X-ray quasi-amorphous materials such as opal A and ferrihydrite.

The authors of this monograph, particularly as teachers and researchers, introduce here some information and comments with regard to the convergences and divergences one actually could face in the literature relative to the definitions of clay and clay minerals.

2.2.1 Historical Retrospective Showing the Lack of Consensus on Clay and Clay Mineral Definitions

A short historical retrospective follows, showing the lack of consensus with regard to clay and clay mineral definitions over the last 70 years.

Grim (1953), a pre-eminent founder of clay science, considers *clay* as:

. . .a rock term and also as a particle-size term; as a rock term, clay is a natural, earthy, fine-grained material which develops plasticity when mixed with a limited amount of water; as a particle-size term, the clay fraction is that fraction composed of the smallest particles,

> which maximum size in geology in the Udden-Wentworth scale is 4 μm and in soil science is the soil separate comprising particles less than 2 μm, according to the Atterberg classification.

Mackenzie (1963) noted the inappropriateness of defining *clay mineral* as "any mineral which occurs in clay" inasmuch as, among several reasons, it would include many accessory minerals that are not characteristic of clay.

Bailey (1980) equated clay minerals with phyllosilicates. However, it is well known that some nonsilicate minerals can impart properties such as plasticity which are characteristic of clays.

Worrall (1982) considers clay as:

> . . . an earth that forms a coherent sticky mass when mixed with water; when wet, this mass is readily mouldable but if dried it becomes hard and brittle and retains its shape; moreover, if heated to redness, it becomes still harder and is no longer susceptible to the action of water; clay is seldom pure, but the substance responsible for its characteristic properties is usually called the clay mineral or clay substance; in addition to the clay mineral, which as was said is the essential and preponderant substance, and so is responsible for the characteristic properties, a number of adventicious minerals, such as quartz, mica and iron oxides are present.

Wilson (1987), a soil scientist, considers that clays are "an important component of soils, where they exert a dominant influence on soil structure and plant nutrition"; also, he considers that one of the fundamental properties of clays is the electric charge on their unit particles, responsible for cations and/or anions sorption, on a reversible basis.

Gomes (1988), as said above, considers various clay concepts depending on the academic education as well as on the technical and professional preparation of those (geologists, pedologists, sedimentologists, mineralogists, petrologists, ceramists, chemists, geotechnists, etc.) interested in this important natural material and resource, either for its genesis, its properties, or for its diverse applications.

Weaver (1989) considers that the term "clay" should only be used in the textural sense to indicate "a natural geomaterial that is finer than 4 μm"; he suggested the term "physils" for the whole family of phyllosilicates (including palygorskite and sepiolite) irrespective of grain size.

Millot (1964,1970, 1989) considers clay as

> . . . a very complex geomaterial; even in the case of being monomineralic clay is composed by a population of distinctive particles, in terms of composition and size, each particle being a population of microdomains; if changes take place in the environment where the particle occurs, each microdomain changes too and, consequently, that particle moves at a proper rate towards a new thermodynamic equilibrium; the dynamics of populations could be also applied to clays.

As mentioned above, in recent years the Nomenclature Committees of AIPEA (Association Internationale Pour l'Étude des Argiles) and of CMS (Clay Minerals Society) have made attempts to achieve the unification of the basic criteria that should be part of clay and clay mineral definitions.

For instance, those nomenclature committees through a joint report by Guggenheim and Martin (1995) introduced the following clay concept: "natural geomaterial essentially composed of fine grained (e.s.d. < 2 μm) minerals, mainly hydrated phyllosilicates, the so-called clay minerals, that exhibits plastic behaviour in the wet state and that gets hard after drying and firing; associated to clay minerals other minerals can occur, the so-called non clay minerals, composing the coarser fractions, but they could be present in the fraction <2 μm too."

One year later Guggenheim and Martin (1996) rationalized and reinforced the concept of clay:

> . . . a naturally occurring material composed primarily of fine-grained minerals, which is generally plastic at appropriate water contents and will harden when dried or fired; although clay usually contains phyllosilicates, it may contain other materials, the so-called "associated phases" that can impart plasticity and harden when dried or fired or may include materials that do not impart plasticity and organic matter.

The term clay mineral refers to "phyllosilicate minerals which impart plasticity to clay and which harden upon drying and firing."

It worthwhile noting that for the first time it is clearly said that plasticity is not a specificity of clay minerals.

According to Harben and Kuzvart (1996) clay means different things to different people, inferring that for the geologist the term "clay" may be used in three senses: (1) to designate particle size <0.002 to 0.004 mm, (2) as a name for a group of minerals known as clay minerals, and (3) as a rock composed essentially of clay minerals.

Moore (1996) and Moore and Reynolds (1997) consider that particle size should be a key parameter in all definitions of clay, and consider that "in geology, sedimentology, and geo-engineering the size limit is commonly set at less than 4 μm e.s.d."

Paquet and Clauer (1997) consider that clay and clay minerals are synonymous terms and distinguish two main types of clays and clay minerals on the basis of the layer charge of clay minerals: "(1) low activity clays or clays of the kaolin group containing phyllosilicates with neutral and slightly hydrated layers; (2) high activity clays or clays of the 2/1 group containing phyllosilicates with charged and more hydrated layers." Other authors have the same understanding about the concepts of clay and clay mineral, that is, they consider both terms synonymous.

Velde and Druc (1999) consider clay as "a natural material composed of about 80% clay minerals and other fine-grained material of different mineral species; this material has a natural grain size of less than 2 μm in diameter." Also, they consider clay minerals, "those particles smaller than 2 μm in diameter which have a specific structure, called a layer or sheet silicate; their grain size and particle structure gives them special physical properties; clay mineral particles form plastic phase in most materials used to make ceramics."

Gomes (2002) distinguishes three concepts: *clay*, *clay mineral*, and *clayey material* or *clay-bearing material*. He considers clay as being "a natural material, mostly composed by fine-grained minerals, mainly hydrated phyllosilicates the so-called clay minerals; clay, in general, exhibits plastic behaviour and hardens after drying or firing; associated to clay minerals other minerals and non-minerals (inorganic and organic cryptocrystalline or amorphous compounds) can occur in clay." Also he considers clay minerals as being "phyllosilicates, essentially hydrous aluminosilicates, fundamental constituents of clay and accessory constituents of clayey rocks and clayey soils, responsible for clay plasticity and hardening after drying and firing." Finally he considers clayey material or clay-bearing material as being "those geologic materials, such as shale, slate, phyllite, marl, marly limestone, lapilli tuff, hyalloclastic tuff, and soils that could bear clay minerals in significant quantity. Despite of some of these materials do not exhibit plasticity in the natural state they could be used, for instance, as ceramic raw materials after being finely ground or blended with other adequate minerals."

Since 2002 *Clay Minerals*, the journal of the European Clay Groups, has changed its name to *Clay Minerals: Journal of Fine Particle Science*. With this title the journal extends the object of clay science to fine-grained minerals and nonminerals that can occur in nature associated with clays other than clay minerals.

Bergaya and Lagaly (2006) consider that the terms "clay" and "clay mineral" are difficult to define, and consider that there is, as yet, no uniform nomenclature for clay and clay mineral.

As a first approximation they consider "clay mineral" as "a class of hydrated phyllosilicates that participate in the fine-grained fraction of rocks, sediments, and soils."

Nonetheless, a consensus about the definition of the clay science objects, clay and clay minerals, is being pursued by the Joint Nomenclature Committees of the AIPEA and CMS.

Distinctive parameters for clay and clay minerals should be taken into account in their concepts, well summarized by Bergaya and Lagaly (in *Handbook of Clay Science*, Elsevier, 2006). For clay: "natural; fine-grained (<2 μm or <4 μm); phyllosilicates as main constituents; plastic; hardens on drying or firing." For clay mineral: "natural and synthetic; any size; may include non-phyllosilicates; plastic; hardens on drying or firing."

Size is not a distinctive property of a mineral species, which is characterized by a specific structure constructed on the basis of certain chemical elements (in some circumstances these basic chemical elements can be isomorphously replaced).

The same mineral species, for instance, quartz, an associated mineral frequently found in clays, can exhibit metric or submicron crystal sizes. It is well established that crystal size in the same clay mineral species is variable. Therefore, size is a parameter or criterion that should not be considered in clay mineral and clay definitions. Considering clay as a fine-grained geological material would be sufficient.

Much of our understanding of clay mineral structures is derived from XRD analysis of macrocrystalline forms of phyllosilicates, such as muscovite, chlorite, and vermiculite, These mineral species regardless of their crystal size are included in clay mineral systematics. However, could a 10 cm × 5 cm × 1 cm crystal of phlogopite or vermiculite that does not occur in any clay deposit be considered a clay mineral?

Meunier (2006) considers that "clay minerals being the most abundant silicates formed at the Earth surface, whatever their origin are always of small grain size." He puts forward the question of why the growth of clay minerals is limited to a few micrometers or nanometers. Based on his experience and on information from published papers he considers that this specificity is related to particular structural aspects (order–disorder and crystal defects).

Any clay has its own *clay fraction* in which the clay particles and clay minerals—the essential constituents of clay—are more concentrated. Conventionally some disciplines set a maximum size for clay particles: pedology <2 μm; geology, sedimentology, and geoengineering <4 μm; colloid science <1 μm.

2.3 Definition of Kaolin and Kaolinite: Inconsistencies Shown in the Literature

2.3.1 Introduction

Concepts can be expressed differently by people from different educational backgrounds and experience. Concepts can also evolve over time.

The concept or definition of kaolin is a good example of such evolution and is of paramount importance for scientific, industrial, commercial, and legal purposes.

Kaolin, one highly valued industrial clay due to its specific functions, for instance, in paper and white ceramics (porcelain, sanitary ware, and porcelain stoneware tiles), could be one such weathering product. If the quantity of kaolin is scarce, for example, not over than 5–8% it would be more appropriate to talk of argillized rock or kaolinized rock instead of kaolin. However, if that quantity, for example, is over 50%, the ultimate product would behave as real clay.

The term "kaolin" is often defined and used differently in science than in industry and commerce. These differences may have legal implications.

In defining kaolin, some questions can be raised. Is kaolin a rock? Is kaolin a mineral? Is kaolin a clay? Is the term kaolin synonymous with the term kaolinite? Is the term kaolin synonymous with the term china clay? What properties must the clay exhibit in order to be called kaolin? Is kaolin only a geological product? Is kaolin a processed or an engineered product too? Is kaolin also a commercial good? Are all kaolins equal? Is the white color of both bulk and processed kaolin a fundamental property? Could other clays rich in kaolin minerals, such as the so-called flint clay, ball clay, and fire clay be considered kaolin as well? What definition of kaolin would meet the joint approval of academia, industry, and the governmental agencies competent to authorize both the exploration and exploitation of kaolin?

Attempts to answer these questions are presented in the following, as far as the aim of this paper is getting fulfilled.

Science privileges the fundamentals of kaolin as a geological product such as genesis, mineralogy, and chemistry.

Industry, on the other hand, privileges the properties of this important industrial mineral and product, such as color, mean grain size and grain size distribution, particle shape, chemical inertia, plasticity, viscosity, brightness, whiteness, opacity, gloss, abrasiveness, and refractoriness. These properties can condition kaolin's manifold uses, markets, and prices.

Kaolin is an industrial product and thus, naturally, a commercial product as well. It is used in the form of certain functional grades, for instance, in ceramics, paper, paint, rubber, fiberglass, medicine, cosmetics, food additives, and agriculture. Commercial grades of kaolin are supplied and transported as dry powder, semi-dry noodle, or as liquid slurry. Recently kaolin has been incorporated into formulations, which as a spray, are applied to fruits and certain vegetables to repel or deter insect damage or to prevent sun scald.

An extensive review of the historical evolution of kaolin definition, from the launching of clay science up to the present time, is presented. This includes the fundamental geological, mineralogical, and technological properties of kaolin as defined by various authors, although not in a well-integrated way.

2.3.2 Background

In regard to kaolin geology, kaolin genesis is diverse and, consequently, the typology of kaolin occurrences/deposits is diverse too. Also, the typology, the volume, and the quality of a kaolin deposit determine the choice of both extractive and industrial processes as well as kaolin economics.

The expressions *kaolin deposit* and *kaolin occurrence* should have different meanings, kaolin deposit being applied only to a kaolin occurrence justified by volume and quality, at both industrial and economic scales, kaolin exploitation, processing, and commercialization.

Kaolin mineralogy and chemistry are fundamentals of kaolin quality, being decisive in terms of kaolin industrial applications and functions. Also their specific properties determine kaolin bulk color.

The white or near-white color in bulk and after drying and firing could be a decisive factor for many important industrial applications of kaolin. However, other types of clay can exhibit a white color (e.g., some deposits of bentonite from Kimolos Island, in the Cyclades, Greece). Even bulk kaolin can exhibit colors other than white or nearly white without being impeditive of certain industrial applications. In mineralogy it is well established that color is not a discriminating property of mineral species. Hence other basic arguments contemplating, for instance, kaolin genesis, mineralogy, and chemistry, common to all types and varieties of kaolin, should be considered.

The emphasis on one or more than one particular property greatly depends on the academic education or professional experience and interest of the individuals seeking a definition of kaolin able to deserve general acceptance.

As a result of all this, a consensual kaolin definition appears to be difficult to achieve.

The volume and quality of a kaolin deposit could be appropriate for its exploitation, but environmental, legal, and administrative factors could hamper such usage.

In Portugal, for example, the concession for exploration and exploitation of kaolin, a highly valuable special clay due to its multiple and specific properties as well as manifold industrial applications, is presently regularized by the so-called Mines' Law issued in 1990. This law is applied to rare mineral resources of high unit value such as gold, cassiterite, massive sulphides, and feldspar/quartz of pegmatite, which are considered to belong to the public domain. The competent authority that approves any proposal submitted for exploration and exploitation of these resources, kaolin included, is the Central Administration through the Department of Energy and Geology with headquarters in Lisbon. Such authority classifies kaolin as industrial rock and as special clay. This same situation has been in force since 1920.

Because in Portugal the adopted kaolin definition considers that raw kaolin is the white or nearly white alteration product of feldspar-rich rocks, this simple fact could be a decisive factor impeding kaolin exploration and exploitation.

On the other hand, the so-called Mineral Masses' law, currently called Quarries' law, governs or regularizes the exploration and exploitation of common mineral resources of low unit value, such as limestone, marble, granite, sand, and common clay, which are considered to belong to the private domain. The exploration and exploitation of these resources, ball clay included, are regularized by the Quarries' law and the competent authority that emits the required license is the local administration, through the county's municipality.

In several countries all mineral resources occurring in, on, or underneath the Earth's surface belong to the state. In China, for instance, a very important country where mineral resources are concerned, both exploration and exploitation of mineral resources depend upon government authorization.

Due to legal regulations not all mineral deposits, including kaolin deposits, despite their interesting reserves and quality, can be exploited, such as mineral resources located inside the so-called protected areas (e.g., agricultural, ecological, or cultural heritage areas), classified on the basis of their unique natural values, such as fauna, flora, agricultural potential, and cultural heritage. When kaolin deposits are not located inside protected areas, and even when the competent exploration and exploitation licenses have been issued, either the potential hazardous environmental impacts raised by nongovernmental environmental groups, or the protest of nearby living populations, could impede any exploration and mining works. And, when justice is involved any decision could be issued so late that the interest in kaolin exploration and exploitation works would disappear.

Kaolin is commonly referred to as "china clay" inasmuch as it was first discovered in China, the term kaolin being derived from *Gaoling* or *Kao-Ling*, meaning "High Hill," a site located near the town of Jingdezhen, in Jiangxi province. Use of the terms *china clay* and *kaolin* is not well defined, but in our opinion similarly to what happened in Kao-Ling, the term "china clay" should be restricted to primary kaolin, the in situ alteration product of the action of hydrothermal and/or weathering processes over feldspar-rich rocks.

All kaolin deposits and the corresponding industrial and commercial products are different. In fact, individual kaolin deposits and natural kaolins vary particularly in many physical properties, which in turn influence kaolin end uses. For many years kaolin's white or nearly white color, both in the raw or bulk state and after firing, was a decisive parameter for kaolin selection in certain industrial applications, particularly in fine ceramics and paper. However, new applications have been found for kaolin, and new advances have been achieved in kaolin processing involving new refining and beneficiation methods, which made possible the use of raw kaolin that exhibits colors other than white. Effectively, kaolin exhibiting yellow, grey, and even other pale colors can be used in many other industrial applications, such as certain types of ceramics (earthenware, stoneware, sanitary ware, as well as wall and floor tiles of earthenware and stoneware), plastics (as filler), cement, paint (as extender), rubber, silico-aluminous refractories, and certain types of pharmaceuticals and cosmetics. Therefore, should the color be a limiting factor in kaolin definition? We do not think so. Properties other than color, such as genesis and composition (mineralogical and chemical), plasticity, and firing behavior should be decisive factors in kaolin definition.

The term kaolin has been applied not only to designate the rock-bearing kaolin minerals, kaolinite and halloysite being the most common kaolin minerals, but also to designate the industrially processed products in which kaolin minerals become concentrated and beneficiated through refining and beneficiation processes, as well as to designate the respective commercial products. However, other names are currently attributed to the rock-bearing kaolin minerals, including kaolinized granite, kaolinized gneiss, kaolinized anorthosite, kaolinized arkose, and kaolinite-bearing sand or clay.

The fundamental properties that make kaolin an industrially important mineral resource, also called industrial mineral, are as follows: chemical inertia within a wide range of pH (3–9), white or near-white color, low electrical and thermal conductivity, easy dispersion in water, low hardness and abrasiveness, high coating capacity when used as pigment, good binder capacity or adhesiveness, and high refractoriness.

Kaolin minerals (kaolinite, halloysite, dickite, and nacrite) are clay minerals; these are fine-grained hydrous phyllosilicates, characterized by planar or nonplanar structures, which are the essential minerals of any clay.

There are no two kaolin exhibiting equal properties. Experience shows that kaolin properties are influenced by the environmental formation and deposition conditions.

A clay-bearing significant amount of kaolin minerals should be called kaolin. Other kaolin minerals-bearing clays, such as ball clay, fire clay, and flint clay, could show similarities to kaolin, in terms of composition (mineralogical and chemical) and color (in the natural state or after firing), although in regard to other properties, for instance, plasticity, viscosity, green-strength, and refractoriness, they could show distinctive behavior. Some types and grades of these sedimentary clays could be called sedimentary kaolin, because the parent or source rocks could be the same as those in the original residual kaolin.

Ball clay deposits, in particular, are less abundant worldwide than kaolin deposits, but ball clay could be as important as standard kaolin in industrial applications.

In general, ball clay shows finer grain size and is of much higher plasticity than standard kaolin; also, its color can range from off-white to black, but after drying and firing, becomes white or near white.

The clays that have determined the origin of the term "ball clay" occur in deposits dating from the Lower Tertiary that fill sedimentary basins in Dorset and Devon, in southwest England. The weathering products from the surrounding deeply weathered uplands of Paleozoic granite and schist were washed and deposited in deltas and lagoons. The layered deposits consist of sand, sandy clay, and clay. Also, clay seams, more or less rich in carbonaceous matter, look brown or black. Kaolinite and illite are the essential clay minerals, but small amounts of interstratified illite–smectite, smectite, and chlorite can also be identified in some clay seams. The distinctive clay mineral assemblages indicate a wide range of ball clay properties, depending on the distinctive contribution of the sedimentation process of the weathering products of both granitic and schistose rocks, the kaolinite structural order being higher or lower, respectively. Therefore, should ball clay deserve to be redefined too? The Al_2O_3 content of some ball clay seams is within the range 30–34%, values typical of high-quality kaolin. Also the K_2O content of ball clay is around 2%, whereas the K_2O content of kaolin is around 1%.

Also, in the Lower Tertiary, sedimentary kaolin of distinctive typology was deposited in tectonic basins of northwestern Portugal, consisting of beds of sand bearing kaolin-rich matrix and ball clay-like beds do occur. In some ball clay seams pseudo-morphs of sillimanite and andalusite have been identified indicating the major contribution of weathered Silurian schistose rocks as source rocks. Also, the kaolinite found in these clay seams is classified as the D (disordered) type.

Ball clay and kaolin are essential components of formulations used in the manufacture of the same products, for instance, tableware, sanitary ware, and wall and floor tiles. Ball clay due to its relatively higher plasticity helps to shape the ceramic bodies, adding green-strength to the body before firing, whereas kaolin provides extra whiteness.

Kaolin could be named special clay, a term attributed to a particular type of clay only produced in a few countries from deposits of small to medium/large size, and characterized as having specific markets and commercially high unit price. The same happens with other kaolin minerals-bearing clays such as ball clay, fire clay, and flint or hard clay, as well as with bentonite, fibrous clays (sepiolite and palygorskite), and Fuller's Earth.

Some treated and modified clays, such as acid-activated and organophyllic bentonite, acid-activated kaolin, and synthetic clays could also be classified as special clays. But the so-called calcined kaolin, the product of heating natural kaolin at high temperatures in a kiln, could not be called kaolin. In fact, calcination alters the size and shape of natural kaolin, increases whiteness and hardness, and improves electrical properties. Calcination at 700–750°C produces metakaolin that can be used either to improve the dielectric properties of PVC cable insulation or used as a pozzolanic concrete additive. The product of kaolin calcination at around 1000°C can be used, especially when coated with silane, to improve both the mechanical properties and the chemical resistance of rubber.

Also, for instance, could ball clay, fire clay (also called *underclay* due to the geological position under coal seams), and flint or hard clay be considered kaolin for scientific, industrial, and legal purposes? In our opinion some particular qualities or grades of these clays, which are rich in kaolin minerals (consequently rich in Al_2O_3, as a rule showing values higher than 30%), and poor in fine-grained staining impurities (iron oxides/hydroxides, titanium oxides, organic matter), should be considered as so. The associated minerals found in these clays are the same as commonly found in kaolin: illite/mica, chlorite, quartz, feldspar, iron oxides/hydroxides (hematite, goethite), titanium oxides (anatase and/or rutile), and organic matter. Therefore, an actual kaolin definition able to integrate the properties common to the above-mentioned geological products and raw materials, the object of many fundamental and applied studies and applications by academic and industrial professionals, would be rather convenient and interesting.

2.3.3 *Retrospective of Kaolin and Kaolinite Definition*

Concepts or definitions are never final. For instance, the definitions of mineral and mineral species have frequently changed during the long history of mineralogy (Povarennykh 1972; Nickel 1995, 1996). Therefore, changes of the kaolin definition have to be considered too, in spite of the relatively short history of clay science, in order to make the kaolin definition compatible with the advances achieved in both fundamental and applied clay science.

According to Ross and Kerr (1931) the name kaolin is a corruption of the Chinese term *kauling* meaning *high ridge*, the name of a hill near Jauchou Fu in China where the material was obtained for the manufacture of whiteware centuries ago. Also, according to the same authors, Johnson and Blake (1867) appear to have first clearly intended the name kaolinite for the "mineral of kaolin".

The first appearance of the term "kauling earth" in a Chinese document was in 1604 AD (i.e., in the middle of the Ming dynasty), and the use of kaolin clay extracted in the Kauling mines located near the town of Chingtehchien, Kiangsi province, for whiteware (mainly porcelain) manufacture, has been documented since the seventeenth century (Chen et al. 1997). These authors consider that the name kaolin would be very likely derived from the name of the actual village Gaoling (Kauling). Also, the authors mention that the term kaolin was introduced in Europe by Père d'Entrecolles, a Jesuit missionary who, in 1712, sent kaolin samples to French authorities (Kuzvart 1977). Kauling mines were abandoned long ago, and presently kaolin is mined in the Dazhou residual deposits about 50 km to the northwest of Gaoling.

Ross and Kerr (1931) defined kaolin as "a rock mass which is composed essentially of a clay material that is low in iron and usually white or nearly white in colour; the kaolin-forming clays are hydrous aluminium silicates of approximately the composition $2H_2O{\cdot}Al_2O_3{\cdot}2SiO_2$ and it is believed that other bases if present represent impurities or adsorbed materials; kaolinite is the mineral that characterises most kaolin."

Grim (1968) defined kaolin as a "fine-grained rock having kaolinite as the main constituent mineral, which containing low iron content shows white or near white colour."

Millot (1964) defined kaolin as "the clay rich in kaolinite, and that the true kaolin deposits are hydrothermal, and that the term kaolin was extended to deposits of sedimentary kaolin and kaolin resulting from atmospheric weathering."

Patterson and Murray (1975) defined kaolin as a "clay consisting of substantially pure kaolinite or related minerals, that is naturally or can be beneficiated to be white or nearly white, will fire white or nearly white and is amenable to beneficiation by known methods to make it suitable for use in white ware, paper, rubber, plastics, paint, ink, and similar uses."

Souza Santos (1975) defined kaolin as the "product of weathering or hydrothermal alteration of feldspar-bearing rocks." The author stated that kaolin could be either formed in situ, the so-called primary or residual kaolin, or transported and redeposited, the so-called secondary or sedimentary kaolin. Also the author stated that in Brazil raw kaolin shows color variations: white, cream, yellow, or rose.

Kuzvart (1977) defined kaolin as a "rock characterized by a useful content of kaolin minerals."

The AIPEA Nomenclature Committee (*in* Bailey 1980), in the classification of phyllosilicates with 1:1 and 2:1 layer types, recommended division into groups on the basis of electrical charge per formula unit, for instance, the group kaolinite-serpentine for 1:1 phyllosilicates. The subdivision of the group into two subgroups: subgroup kaolinite and subgroup serpentine, on the basis of dioctahedral and trioctahedral characteristics, has also been recommended.

Worrall (1982) defined kaolin "as one of two main groups of clay mineral or clays: kaolins and montmorillonites." For the author the chemical composition of these clays was coincident with the chemical composition of their main representative clay minerals, kaolinite and montmorillonite. Also, the author classified clays into two types—residual and sedimentary—and considered other clays, ball clay and fire clay, close to kaolin (china clay) in chemical composition, but differing a great deal in other properties and application fields.

Wilson (1987) defined kaolin as "the clay which consists principally of the clay mineral kaolinite."

López-Colmenar (1992) defined kaolin as "all massive rock containing a variable percentage of clay minerals of composition equal or approximate to the composition of kaolinite – $2SiO_2{\cdot}Al_2O_3{\cdot}2H_2O$ – easy to concentrate through the separation of the associated minerals"; in the case of its being a clayey rock, making the concentration of the hydrous aluminosilicates difficult and expensive, it is called *kaoliniferous* or *refractory clay* in Spain.

Manning (1995) considered kaolin (or china clay), ball clay, and bentonite as the "dominant industrial clays, due to their diverse and very important industrial applications." The author defined kaolin as "chemically inert material that can be prepared as a white powder specified (in part) according to its whiteness and brightness; in terms of genesis the author distinguished primary and secondary kaolinite deposits, including kaolin (china clay) in the primary deposits, and ball clay in the secondary deposits".

Guggenheim and Martin (1995) made a clear distinction between clay and clay mineral. These authors defined "clay as a natural geologic material mainly composed of fine-grained (less than 2 μm) minerals essentially hydrous phyllosilicates, the so-called clay minerals which, in general, show plastic behaviour and that harden after drying or firing; associated to clay minerals other minerals can occur, the so-called non clay minerals, more concentrated in coarse fractions, but that could be still present in the less than 2 μm fraction." Also, the same authors defined "clay minerals as phyllosilicates, essentially hydrous aluminosilicates, which are essential components of clays and accessory components of many other rocks, being responsible for clay plasticity and for its hardening after drying or firing."

The AIPEA Nomenclature Committee (Guggenheim and Martin 1996) recommended that the term "kaolin-serpentine" should be used instead of the term "kaolinite-serpentine."

According to Moore (1996) the term "clay" has been used as a mineral term, as well as a size term and a rock term. Hence the term "kaolin" could be applied to clay and to clay mineral.

Paquet and Clauer (1997) classified clays into three groups: *low activity clays, high activity clays,* and *intermediate activity clays*. This classification was based on the electrical charge (z) of the structural unit cell, and they defined kaolin, represented principally by kaolinite, "as low-activity clay, having a low layer charge, nearly zero."

Harben (1999) defined kaolin "as white, soft and plastic clay predominantly composed of fine-grained platy kaolinite." As regards the color, the author states that kaolin can show white color, sometimes with red, brown, yellow, and blue tints. Also, the same author classified kaolin as a clay belonging to the group of kaolinite-bearing clays, a group that in addition to kaolin itself includes other special clays also predominantly composed of kaolinite, such as ball clay (a fine-grained, highly plastic kaolinite admixed with quartz, mica, illite, smectite, chlorite, and colloidal carbonaceous matter); flint clay (a smooth, microcrystalline, hard-slacking rock with a pronounced conchoidal fracture, composed mainly of kaolinite, which becomes plastic when ground in water); and fire clay (a silico-aluminous clay composed of a disordered variety of kaolinite capable of withstanding high temperatures without deforming).

The halloysite-bearing clay, called *halloysite clay*, is also classified as kaolin, being predominantly composed of halloysite, a clay mineral similar to kaolinite in terms of chemical composition and structure, but whereas halloysite particles can exhibit platy, rolled, or spheroidal shapes, kaolinite particles exhibit a platy shape.

Harvey (2000), based on studies carried out on industrial clays from various worldwide deposits, classified such clays into four categories (three for special clays, one for common clays), depending upon the sophistication and economics of the products manufactured with them, and the sophistication of the technologies required for their development. In this classification kaolin could be included in the three categories of special clays.

The first category includes high-quality kaolin, whose large-scale production and processing require advanced and expensive technologies, and consequently high investment; this type of kaolin has a large circulation in both national and international markets, being represented, for instance, by kaolin from Cornwall, United Kingdom, by kaolin from Georgia, United States, and by kaolin from Amazonia, in Brazil.

The second category also includes high-quality kaolin whose production and processing, although on a small scale, requires high technologies; this type of kaolin is addressed to small markets, both national and international, and represented, for example, by kaolin mostly consisting of halloysite from Maugaparerua, Matauri Bay and Mahimali, in New Zealand. The third category includes medium-quality kaolin, whose production and processing requires medium technologies and investment, represented, for example, by central and western European kaolin, from Germany, Czech Republic, Ukraine, France, Spain, and Portugal.

Galán (2003) distinguished *commercial clays* from *noncommercial clays*, the former being those with possible industrial application. The author defined "commercial kaolin as a rock that contains a certain proportion of minerals of kaolin group (kaolinite, dickite, nacrite and halloysite), and that could be economically extracted and concentrated." The author includes in commercial kaolin,

> the kaolinite-bearing clays (ball clay, fire-clay and flint clay) predominantly composed of kaolin minerals; raw kaolin, in general, is not a clay, but a rock (kaoliniferous sand or arkose, and kaolinized granite, gneiss, rhyolite or anorthosite), reason why it needs to be processed in order to concentrate kaolin minerals; on the contrary, kaolinite-bearing clays, which in the natural state are, in general, richer in kaolin minerals than kaolin are not processed, requiring only milling, sieving, atomization, etc.

With regard to the distinction between *clay* and *clay mineral*, in 2006 the actual position of the JNCs (Joint Nomenclature Committees) of both the AIPEA and the CMS on using *clay as a rock term* was still indeterminate. Nevertheless, the literature often uses the term "clay" for clay mineral, for example, kaolin instead of kaolinite, because the former name is shorter and less cumbersome.

Bergaya and Lagaly (2006) established the distinction between *clay* and *clay mineral* based on the following parameters: genesis, granularity, mineralogy, plasticity, and behavior on drying or firing. Clay is natural, fine-grained (<2 or <4 μm), mainly constituted of phyllosilicates, plastic (except flint clays), and hardens on drying or firing. On the other hand, clay mineral is natural or synthetic, without size criterion, might include nonphyllosilicates, plastic, and hardens on drying or firing.

Such concepts could be extended to the clay kaolin and to the clay mineral kaolinite.

2.3.4 Discussion

From the information above and from the evolution of kaolin definition over the relatively short history of clay science the unconformity between the arguments found relevant for academic and industrial scientists in what concerns kaolin definition becomes clear.

Primary or residual kaolin and secondary or sedimentary kaolin are distinctive in terms of genesis, occurrence, and properties, and consequently in terms of industrial processing methodologies, uses, and functions.

Frequently, ball clay, flint clay, and fire clay, while in the bulk state, contain higher contents of kaolinite and/or halloysite than bulk kaolin, although after being refined, certain size fractions of kaolin, compared to equivalent size fractions of ball clay and fire clay, contain higher concentrations of kaolinite and/or halloysite.

Fire clays have very distinctive properties relative to ball clays. Fire clays occur underneath coal seams from the Carboniferous, the reason why they are also called *underclays*. Certain fire clays from Yorkshire, in England, contain dickite as an important constituent associated with kaolinite (Gomes 1979), and the Al_2O_3 content can reach values slightly above 30%.

In terms of industrial applications ball clay is more important than fire clay, despite the participation of this clay both in silico-aluminous refractories and in stoneware. As a matter of fact, ball clay is an essential constituent of ceramic pastes used for the production of certain types of porcelain, sanitary ware, earthenware, and stoneware (household ware and floor tiles of "porcellanic stoneware" or "porcelanato").

Considered as secondary kaolin, ball clay is geologically associated with primary or residual kaolin. In southwest England Lower Tertiary ball clay deposits occur in small or medium sedimentary basins located near the kaolin deposits derived from the alteration of Hercynian granite outcroppings in the counties of Devon and Cornwall.

English ball clay is classified into groups according to their main properties: chemical and mineralogical composition, content of carbonaceous matter, grain size distribution, <2 μm content, unfired brightness, fired brightness, and viscosity concentration.

High grades of ball clay contain: 65–75% kaolinite (60–70%); 15–20% mica/illite (15–20%); 30–34% Al_2O_3 (28–32%); 1–2% Fe_2O_3 (1–1.5%); 2–3% C (0.1%); and <2 μm 75–90% (65–70%). Inside the parentheses are the corresponding contents of medium-grade ceramic kaolin.

Mitchell and Vincent (1997) considered four types of ball clay. Three of these types are distinguished by their multilayered structure, thickness of ball clay layers, and commercial value. Type 1 of ball clay is represented, for instance, by deposits in the Bovey and Petrockstow basins in England, dated Lower Tertiary. Types 2 and 3 of ball clay are represented by the deposits in Westerwald in Germany, by the deposits in Skalma in the Czech Republic, by the deposits in Tennessee and Jackson/Purchase in the United States of America, and the deposits in Dorset in England, respectively, all dated Lower Tertiary as well. Type 4 ball clay corresponds to deposits of small to medium dimension and relatively small thickness, formed in lacustrine and lagoonar environments, and dated Pliocene–Pleistocene.

Kaolin definition is as difficult as clay definition. Indeed, clay definition differs a great deal among people having diverse educational and professional experience (Gomes 2002).

Similarly kaolin definition could show differences according to the educational or professional background of people interested in this special clay. The only disagreement we have with regard to the aforementioned definitions of clay and clay mineral by Bergaya and Lagaly (2006) is the attribution of clay mineral to either natural or synthetic products.

The designation *clay mineral* should be reserved only for natural, inorganic, crystalline, hydrous phyllosilicates. Synthetic or artificial inorganic, crystalline, hydrous phyllosilicates that could be produced in the laboratory, despite being analogues of clay minerals, should not be considered as such.

The AIPEA Nomenclature Committee (Guggenheim et al. 1997) endorses the use of both commercial names and synthetic names for the nomenclature established by the Commission on New Minerals and Mineral Names (CNMMN; Nickel 1996). Therefore, synthetic minerals should be referenced by enclosing the name in quotation marks, unless the synthetic mineral has been recorded as a trademark, in which case the name starts with a capital letter.

Trying to integrate all that was said before we consider kaolin as:

> a geologic material that is the outcome product of the alteration of rocks bearing significant contents of feldspars (essentially aluminosilicates of potassium or sodium and calcium) and micas (essentially aluminosilicates of potassium), alteration due to either weathering or deuteric (essentially hydrothermal) processes, or due to both processes too. Those rocks could belong to distinct genetic types: igneous rocks, such as, rhyolite, granite, granodiorite and anorthosite; metamorphic rocks, such as, gneiss, migmatite and micaschist; or sedimentary rocks, such as, feldspar-bearing sandstone (named arkose) and quartz sand containing some feldspar, mica and a clayey matrix that could exhibit white, grey, and yellow colours. The term kaolin would be attributed to the kaolinite-bearing rock as well as to the products that would result from its refining and beneficiation in order to be commercialized for specific industrial applications.

In clay mineralogy, taking into account the recent recommendations of the AIPEA Nomenclature Committee (Guggenheim et al. 2006), the expression *kaolin minerals* is applied to the subgroup *kaolin minerals* which with the subgroup *serpentine minerals* constitutes the group *kaolin-serpentine minerals*.

Kaolin minerals are classified as hydrous phyllosilicates having planar structures (platy crystal shapes) based on a layer type 1:1 of dioctahedral character, and comprising the following minerals: kaolinite, dickite, nacrite, and halloysite (in the case where it has planar structure), the first three minerals having the same chemical composition but distinctive structures called polytypes.

Halloysite, in addition to the planar structure, can also exhibit nonplanar structures (rolled or spheroidal shapes), and differs from kaolinite because it also has the same amount of structural water as it has constitution water.

There are kaolin deposits in which kaolinite is the predominating kaolin mineral, represented, for instance, by Seilitz deposits of residual type, in Germany, by St. Austell deposits of mixed hydrothermal/residual type, in Cornwall, England, by Georgia deposits of sedimentary type in the United States, and by Rio Jari (tributary of the Amazon river) and Rio Capim deposits of sedimentary type in Brazil.

Other kaolin deposits exist in which halloysite is the predominating kaolin mineral represented, for instance, by Maungaparerua, Matauri Bay, and Mahimali deposits of hydrothermal/residual type in New Zealand. Also, there are kaolin deposits, represented by Suzhou and Dazhou kaolin deposits of hydrothermal/residual type in China, in which kaolinite and halloysite occur in similar contents.

Finally, there are kaolin deposits of hydrothermal origin, well represented, for example, in Guanajuato, São Luis Potosi, and Nayarit deposits in Mexico, in which dickite and nacrite are the predominating kaolin minerals. In Tragos (Mangualde, Portugal) there is a small but interesting kaolin occurrence of hydrothermal type related to uranium mineralization, in which dickite was the only kaolin mineral identified as occurring in relatively long and perfect pseudo-hexagonal crystals coating tiny bipyramidal quartz observable only through a microscope (Gomes 1979).

Kaolin and ball clay are mainly used in ceramics, rubber, plastics, and agrochemicals.

Kaolin can occur in distinctive geological environments, either as residual deposits, or as redeposited or sedimentary deposits, as with any other special clay or common clay. Kaolin from one deposit has specific characteristics or properties (color, chemistry, mineralogy, clay fraction content, grain size and grain size distribution, plasticity, and rheology) that differ from the corresponding properties of kaolin from any other deposit. Even in the same deposit the heterogeneity in terms of kaolin characteristics is a common occurrence, and within kaolin properties color variations from white to grey, yellow, brown, and red could be easily observable in the field. Therefore, the results of studies carried out in the kaolin from a certain deposit cannot be extrapolated to any other kaolin deposit.

The paper by Murray and Keller (1993) entitled "Kaolins, Kaolins, Kaolins" is a clear indication that all kaolins are not equal, being distinctive in terms of genesis, composition, and properties. Individual kaolins vary particularly in many physical characteristics (e.g., particle size, particle size distribution, particle shape, particle aspect ratio, viscosity, brightness, whiteness, abrasiveness, opacity, and gloss), which in turn influence their end uses.

Is kaolin a mineral that is a natural (formed as a result of geological processes), homogeneous inorganic solid, normally crystalline? The answer is negative although, presently, some researchers consider the term kaolin equivalent to the term kaolinite, a clay mineral commonly found in kaolin that is a real mineral species represented by the chemical formula $Al_2Si_2O_5(OH)_4$. Experience shows that kaolin is an association of clay minerals (essentially belonging to the kaolin subgroup of minerals that comprises the mineral species, kaolinite, halloysite, dickite, and nacrite), nonclay minerals (quartz, feldspar, mica, and iron oxides/hydroxides), and sometimes noncrystalline phases (organic matter, and noncrystalline iron or manganese hydroxides).

Could we call kaolin a rock that by definition is a natural consolidated or unconsolidated aggregate or association of minerals and, within these, the so-called kaolin minerals? Yes, it could be in particular cases when kaolin subgroup minerals are predominating, even when the structure of the parental rock (e.g., granite or arkose) is visible. In this situation, for instance, the expressions kaolinized granite or kaolinized arkose are also currently used, and such expressions with good reason would be applied if kaolin minerals do not predominate.

However, could the term kaolin be used if the kaolin minerals content of the geological material is low? Yes, it could, because we know that kaolin mineral content around 8–10%, constituting the white or nearly white matrix of sand and low consolidated sandstone deposits, could be sufficient for the economic extraction and concentration of both the fine fraction and the coarse fraction to be feasible, particularly if the coarse fraction (sand and pebble) could also be commercialized for ceramics or civil construction works after being submitted to appropriate washing and classification.

The designation "kaolin" attributed to the fine fraction (<0.074 mm or even finer) is quite acceptable in mineralogical, industrial, and commercial terms.

On the other hand, could the term "kaolin" be attributed to the fine white powder mainly consisting of kaolinite or halloysite identified by X-ray diffraction, and that partially replaces the individual feldspar or mica crystals which constitute a certain rock, such as granite, granodiorite, anorthosite, arkosic sandstone, and gneiss? No, if the alteration product could not be industrially and economically concentrated in order to be commercialized. Or could kaolin be the name of the very scarce white powder present in the interior of feldspar crystals in apparently nonaltered or little altered granite, powder that is just observable through an optical microscope or after granite fragmentation and washing, and identified by XRD as being mainly composed of kaolinite? The answer is negative as well.

On the other hand, the reddish kaolinite-bearing clay size component of a ferrallitic soil that could be relatively abundant in some soils of this typology should not be called kaolin. This type of soil, also called lateritic soil, is used as construction material in civil works, for instance, bricks for housing and road pavement, should not be called kaolin.

In this specific case the clay size component, in spite of its reddish color, could be turned into white or nearly white color on a laboratory scale by dissolution of the iron-staining pigment.

The three mineral products referred to, despite the existence in them of the minerals characteristic of kaolin, should not be classified as kaolin; it is not only the scale (micro, meso, or mega) of the occurrences that matters, but also the feasibility and economics of the industrial processing as well as the existence of markets for the end products.

As the outcome of the preceding discussion the kaolin designation should contemplate and satisfy not only certain mineralogical requisites, but also industrial and commercial requisites that depend on specific kaolin properties, such as color, brightness, mean particle size, particle size distribution, particle shape, viscosity, plasticity, abrasiveness, and refractoriness; a particular kaolin has its proper identity expressed by specific properties.

Up to the 1960s of the last century the main industrial use of kaolin was in ceramics to produce diverse types of whiteware ceramics, such as porcelain, earthenware, sanitary ware, stoneware, and bone china, the last belonging to the phosphatic type of whiteware, whereas the others belong to the feldspathic type of whiteware. At that time kaolin was defined as white or nearly white clay that after being dried and fired gets increasingly hard maintaining the white color.

The referred products require kaolin of distinct brightness; for instance, porcelain and bone china require kaolin of higher brightness grades than the other products. Many of these products (household porcelain, art porcelain, vitreous porcelain, electrical porcelain, and cookware porcelain) are glazed using adequate porcelain glazes, some more opaque than others, glazes that could hide the color base of the ceramic body. Therefore, distinctive kaolin brightness grades for distinct functions have been established, corresponding to their distinctive unit prices. Thereafter, for about 30 years kaolin became a fundamental raw material for the production of certain types of white paper, incorporated as a white pigment applied either as coating or as filler. Kaolin grades characterized by higher brightness, better perfection of particle shape, and lower abrasiveness are required more for coating grades rather than filler grades.

Kaolin is characterized in general by its current white color, fine granularity, low plasticity, and low viscosity compared to other special clays. Similar to any other clay, kaolin is essentially formed and occurs in surface environments of the Earth.

Raw kaolin usually shows white color, its fundamental property when fabricating products that require raw materials with high brightness, such as porcelain and high-quality printing paper. However, in its natural state kaolin frequently can show colors with tints of yellow, red, and grey, that in the first two cases is attributed to cryptocrystalline, weakly crystalline, or crystalline iron oxides/hydroxides, and in the third case is attributed to the presence of finely divided organic matter.

None of these color tints hinders kaolin use in equally important applications such as earthenware, sanitary ware, stoneware, rubber, thermoplastics, and certain types of cement, mortar, pozzolana, and refractories. During the thermal processing required for the manufacture of these products, the yellow and grey tints in particular change, depending upon the kiln atmosphere (reducing or oxidizing), into white or near white tints.

On the other hand, there are beneficiation processes of current applications for kaolin bleaching, for instance, chemical dissolution, selective flocculation, ultraflotation, froth flotation, delamination, and high-intensity magnetic separation, processes whose application intends significantly to reduce or eliminate certain contaminants or penalizers, such as iron oxides/hydroxides, titanium oxides, biotite, and tourmaline.

Exceptionally kaolin does not require refining and beneficiation, such as the kaolin from Mevaiela, Angola, that is the product of anorthosite alteration (Gomes et al. 1994). The chemical composition of raw kaolin and of its <0.038 and <0.002 mm fractions is almost invariable.

Kaolin in a particular geological deposit can exhibit chromatic variations, either laterally or vertically, and the kaolin showing these color variations could be selectively exploited and processed in order for the output of specific commercial grades for certain industrial applications and functions to be feasible. As a matter of fact, color is not a distinctive property of kaolin. Other special clays rich in kaolinite, such as flint clay, ball clay, and fire clay, can show in the natural state, or after firing, white or nearly white color.

Even special clays in the natural state not bearing kaolinite, such as fibrous clays (sepiolite and palygorskite), and bentonite and ball clay can show white or near-white color. These chromatic variations are related to the above-mentioned contaminants and, as a rule, they are more common in secondary than in primary kaolin deposits.

In actuality, the most distinctive properties of kaolin relative to other special clays (ball clay, flint clay, and fire clay) are their mineral and chemical composition and the grain size distribution.

Kaolin fine fractions (<0.063, <0.053, <0.045, and <0.010 mm), potentially interesting for specific industrial applications, are mainly composed of the clay minerals kaolinite $Al_2Si_2O_5(OH)_4$, and/or halloysite $Al_2Si_2O_5(OH)_4$, $2H_2O$. Illite, chlorite, and smectite can occur as associated accessory clay minerals. Quartz, mica, feldspar, and oxides/hydroxides of Fe, Al, and Mn are other associated minerals that can occur. Also, as associated phases, organic matter and other noncrystalline compounds can occur as well.

Kaolin being special clay occurs in deposits that, not being frequent worldwide in the geologic environment, should be of medium to high dimension and volume in order to allow its economic exploitation. Kaolin mining and technological processing need extractive, refining, and beneficiation methods that require high financial investment and sophisticated technological processes.

The final product of kaolin processing will have a higher or lower unit value, depending upon either kaolin-specific natural characteristics, or the equally specific properties of the final refined and beneficiated product. Essentially, through the application of processes of hydraulic or aero-flotation treatment to the so-called kaolin parent rock, refining promotes the concentration of finest particles, and consequently the concentration of the clay minerals characteristic of kaolin. Therefore, kaolin, having specific granulometric characteristics (grain size less than 0.045 mm or less than 0.010 mm) and other specific properties, can be produced and commercialized.

Refining modifies the granulometric or textural composition of kaolin, and also modifies properties such as mineralogical and chemical composition. On the other hand, beneficiation promotes the modification of properties such as color, brightness, gloss, and rheology.

To be feasible, in economic terms, the exploitation of a kaolin deposit, the minimum content of the industrially processed and commercialized kaolin fraction, that is, the size fraction showing white or other colors showing grey, yellow, and other tints, should be within the range 8–10%.

Kaolin is, comparatively, one of the most important mineral resources in economic terms, due to its diversity of application fields and functions, such as coating and filler in paper, as an essential component of ceramic pastes to produce certain ceramics (diverse types of porcelain, earthenware, sanitary ware, and stoneware), as filler in rubber and plastics, and as important raw material to produce certain types of paint, glass fiber, cement, mortar, mortar, pozzolana, refractories, pharmaceuticals, cosmetics, and the like.

As said before, in terms of genesis and geologic occurrence, kaolin is classified into four types: residual, hydrothermal, residual-hydrothermal, and sedimentary, the first three included in the group of primary kaolin and the fourth in the group of secondary kaolin.

2.3.5 Conclusions

From the aforementioned information, and the questions and proposed answers, some conclusions may be drawn:

1. Kaolin in kaolin deposits, as a rule, is heterogeneous in terms of many relevant properties required for industrial applications, such as: (a) color of the kaolin-bearing materials shows changes, either laterally and vertically; (b) textural features of kaolin-bearing materials show changes, either laterally or vertically; (c) mineralogy and chemistry of both bulk kaolin and their size-fractions show changes, either laterally or vertically; (d) kaolin mineral quality and quantity, fundamental parameters in terms of deposit economics, show changes, either laterally or vertically.
2. In the same kaolin deposit kaolin-bearing materials having distinctive origin or genesis could occur in association; for instance, in a certain kaolin deposit, residual type kaolin could be overlaid by sedimentary type kaolin, and this last type of kaolin could show layers of diverse textural facies, kaolinitic clay, or fine sand and coarse sand both bearing kaoliniferous matrix.
3. Color of natural kaolin, despite being an important property for some industrial applications and functions, is not an impeditive factor of kaolin use for other important applications; raw kaolin colors other than white, such as beige, pink, yellow, and grey can be attenuated, submitting the natural refined kaolin either to adequate beneficiation treatment, or to adequate thermal treatment carried out during the transforming process of common application, for instance, in ceramics, refractories, cements, pozzolana, and metakaolin.
4. Color is not a necessary requisite to be included in a kaolin definition, as well as other properties, such as plasticity and hardening after drying or firing, because they are not exclusive to kaolin, and in addition, they are not common to all kaolins. For instance, flint or hard clay, in the natural state, is already very consolidated and hard, and is nonplastic unless after being finely ground.

5. Certain types of ball clay, an industrial clay having important applications in common with kaolin, in the case of showing high kaolinite content, high alumina content, low iron content, and near-white color after firing, should be considered sedimentary kaolin for scientific, industrial, and legal purposes.
6. The presence of kaolinite (or its polytypes, dickite and nacrite) and/or halloysite, either in the clay-bearing or argillized rock, or in its processed and commercial products, is a fundamental requisite for kaolin classification.

A proposal for a new kaolin definition, characterized by its more comprehensive character able to integrate its geology (origin and occurrence), mineralogy, technological properties, as well as the always decisive economic and legal criteria, is put forward:

> Kaolin is a fine-grained natural or engineered product predominantly composed of clay minerals classified into the kaolin subgroup minerals; to fulfill the diverse and important industrial functions kaolin is extracted from geologic deposits of previously assessed quality, economics, and legality; to be commercialized kaolin could undergo appropriate refining and beneficiation in order to achieve its required ultimate white or near-white color and other relevant properties.

In conformity with this definition the term "kaolin" could be applied not only to the special kaolin-bearing clay currently called kaolin, but also to other special kaolin-bearing clays, particularly high-quality ball clay and flint or hard clay. Therefore, for scientific, industrial, commercial, and legal purposes, ball clay and flint clay deposits whose high quality and volume would have been recognized by the competent administrative entity, based upon adequate field and laboratory investigation, and conformity to national legislation, should deserve the same classification and treatment as kaolin deposits.

2.4 Evolution of the Definition of Illite

The concept of illite has also changed with time. Grim et al. (1937) proposed the term *illite* as "a general term, not as a specific clay mineral name, for the mica-like clay minerals." The term *illite* includes both di-and trioctahedral structures, and has been used worldwide for a long time as a name for clay minerals with $d_{(001)} = 10$ Å spacing that shows substantially no expanding-lattice characteristics, that is, no swelling clay components.

Bailey et al. (1984) in the report of the Clay Minerals Society Nomenclature Committee (in *Clay Minerals*, 32:239) have proposed a new definition for illite: "a non-expandable dioctahedral clay mineral; ionic substitutions in both octahedral and tetrahedral sites of the mica structure; contains a phengite component; the interlayer substitutions can include not only K^+ but also H_3O^+ ions; and an interlayer charge ranging between 0.8 and 0.6."

Meunier and Velde (2004) following the investigation of illite from distinctive natural systems (soil and other geological systems) and processes (diagenesis, supergene, and hydrothermal alteration) redefined illite as: "a low-potassium mica (i.e., with a 10 Å basal spacing); able to undergo crystal growth (what is proper of a mineral species); with a K content per unit cell near 0.9 and Mg and Fe contents depending upon the chemical system in which illite is formed; and that could occur as 1M and 2 M polytypes."

Martin et al. (1991) in the report of the Clay Minerals Society Nomenclature Committee classify the hydrous phyllosilicates structure into two types: planar and nonplanar, illite being characterized by its planar structure.

Recently Guggenheim et al. (2006) in the report of the Nomenclature Committee for 2006 of AIPEA maintain the classification of hydrous phyllosilicates into two types: planar and nonplanar or modulated, and defined illite as "planar hydrous phyllosilicate that belongs to the interlayer-deficient mica group, and has a dioctahedral character. In the classification of planar hydrous phyllosilicates illite is considered as being an interlayer-deficient mica having non-hydrated mono-or divalent cations (~0.60–0.85)."

References

Bailey SW (1980) Summary of recommendations of AIPEA nomenclature committee. Clay Clay Miner 28:73–78; Clay Miner 15:85–93

Bailey SW (1982) Nomenclature for regular interstratifications, report of the clay minerals society nomenclature committee for 1980–1981. Clay Clay Miner 30:76–78; Report of the AIPEA Nomenclature Committee, AIPEA Newsletter No 18, Supplement of February, 1982

Bailey SW (1989) Report of the AIPEA nomenclature committee. AIPEA Newslett 26:17–18

Bailey SW, Brindley GW, Johns WD, Martin RT, Ross M (1971a) Summary of national and international recommendations on clay mineral nomenclature, 1969–70, CMS nomenclature committee. Clay Clay Miner 19:129–132

Bailey SW, Brindley GW, Johns WD, Martin RT, Ross M (1971b) Clay mineral society report of nomenclature committee, 1969–70. Clay Clay Miner 19:132–133

Bailey SW, Brindley GW, Kodama H, Martin RT (1979) Report of the clay minerals society nomenclature committee for 1977 and 1978. Clay Clay Miner 27:238–239

Bailey SW, Brindley GW, Fanning DS, Kodama H, Martin RT (1984) Report of the clay minerals society nomenclature committee for 1982 and 1983. Clay Clay Miner 32:239

Bailey SW, Alietti A, Formoso MLL, Konta J, Koster HM, Mackenzie R, Morgan DJ, Mumpton FA, Nagasawa K, Paquet H, Raussell-Colom JA, Zvyagin BB (1986) Report of the AIPEA nomenclature committee (Illite-Glauconite-Volkonskoite). AIPEA Newsletter No 22, Supplement of February

Bergaya F, Lagaly G (2006) General introduction: clays, clay minerals, and clay science. In: Bergaya F, Theng BKG, Lagaly G (eds) Handbook of clay science, Developments in clay science. Elsevier, New York, 1224 pp

Brindley GW, Pedro G (1976) Meeting of the nomenclature committee of AIPEA, Mexico City, July 21, 1975. AIPEA Newslett 12:5–6

Brindley GW, Pedro G (1970) Report of the AIPEA nomenclature committee. AIPEA Newslett 4:3–4

Brindley GW, Pedro G (1972) Report of the AIPEA nomenclature committee. AIPEA Newslett 7:8–13

Brindley GW, Bailey SW, Faust GT, Sa F, Rich CI (1968) Report of the nomenclature committee (1966–67) of the clay minerals society. Clay Clay Miner 16:322–324

Chen P-Y, Lin M-L, Zheng Z (1997) On the origin of the name kaolin and the kaolin deposits of the Kauling and Dazhou areas, Kiangsi, China. Appl Clay Sci 12:1–25
Galán E (2003) Arcillas especiales I: caolines y arcillas caoliniferas. In: Galán E (ed) Mineralogía aplicada. Editorial Síntesis, Madrid, pp 105–135
Gomes CSF (1979) Characterization of structural defects in kaolinite minerals, PhD thesis, Department of Ceramics, University of Leeds, United Kingdom, 240 pp
Gomes CSF (1988) Argilas: O que são e para que servem, Fundação Calouste Gulbenkian (ed), 457 pp
Gomes CSF (2002) Argilas: Aplicações na Indústria. In: Gomes C (ed), Universidade de Aveiro, 338 pp
Gomes CSF, Worrall WE (1979) A dickite showing unusual crystal habit and structural disorder. In: Mineralogy of ceramics. Proceedings of the British Ceramic Society, vol 28, pp 21–105
Gomes CSF, Velho J, Guimarães F (1994) Kaolin deposit of Mevaiela (Angola), alteration product of anorthosite: assessment of kaolin potentialities for applications in paper. Appl Clay Sci 9:97–106
Grim RE (1953) Clay mineralogy. McGraw-Hill, New York, 384 pp
Grim RE (1968) Clay mineralogy, geological science series. McGraw-Hill, New York, p 29, 569 pp
Grim RE, Bray RM, Bradley WF (1937) The mica in argillaceous sediments. Am Mineral 22:813–829
Guggenheim S, Martin RT (1995) Definition of clay and clay mineral: Joint report of the AIPEA nomenclature and CMS nomenclature committees. Clay Clay Miner 1995(43):255–256; Clay Miner 30:257–259
Guggenheim S, Martin RT (1996) Reply to the comment by more DM on "definition of clay and clay mineral", Joint report of the AIPEA and CMS nomenclature committees. Clay Clay Miner 44:713–715
Guggenheim S, Alietti A, Drits VA, Formoso MLL, Galán E, Köster HM, Paquet H, Watanabe T, Bain DC, Hudnall WH (1997) Report of the association internationale pour l'Étude des argiles (AIPEA) nomenclature committee for 1996. Clay Clay Miner 45:298–300; Clay Miner 32:493–496
Guggenheim S, Bain D, Bergaya F, Brigatti MF, Drits V, Eberl DD, Formoso M, Galán E, Merriman RJ, Peacor DR, Stanjek H, Watanabe T (2002) Report of the AIPEA nomenclature committee for 2001: order, disorder, and crystallinity in phyllosilicates and the use of the "crystallinity index". Clay Clay Miner 50:406–409; Clay Miner 37:389–393
Guggenheim S, Adams JM, Bain DC, Bergaya F, Brigatti MF, Drits VA, Formoso MLL, Galán E, Kogure T, Stanjek H (2006) Summary of recommendations of nomenclature committees relevant to clay mineralogy: report of the association internationale pour l'Étude des argiles (AIPEA) nomenclature committee for 2006. Clay Clay Miner 54:761–772
Harben PW (1999) The industrial minerals handbook, 3rd edn. IML, Tonbridge, 296 pp
Harben PW, Kuzvart M (1996) Industrial minerals: a global geology. Industrial minerals information. Metal Bulletin PLC, London
Harvey CC (2000) Categorization and economic assessment of industrial clays, the clay minerals society workshop on "industrial clay mineralogy". Loyola University, Chicago
Johnson SW, Blake JM (1867) On kaolinite and pholerite. Am J Sci 43:351–361
Kuzvart M (1977) Aspects of kaolin genesis. In: Galán E (ed) Proceeding of the 8th kaolin symposium and meeting on alunite. Ministerio de Industria y Energia, Madrid-Rome, Madrid, p K-12
López-Colmenar JMM (1992) Caolin y arcillas caoliniferas, recursos minerales de españa. In: Guinea JG, Frias JM (ed) Textos Universitarios, Consejo Superior de Investigaciones Científicas, pp 27–44
Mackenzie R (1963) De natura lutorum, clays and clay minerals. Proceedings of the eleventh national conference on clays and clay minerals, vol XI, Pergamon Press, Oxford, pp 11–28
Manning DAC (1995) Introduction to industrial minerals. Chapman & Hill, London, 276 pp

Martin RT, Bailey SW, Eberl DD, Fanning DS, Guggenheim S, Kodama H, Pevear DG, Srodon J, Wicks FJ (1991) Report of the clay minerals society nomenclature committee: revised classification of clay minerals. Clay Clay Miner 39:333–335
Meunier A (2006) Why are clay minerals small? Clay Miner 41(2):551–566
Meunier A, Velde B (2004) Illite. Springer, Berlin, 286 pp
Millot G (1964) Géologie des argiles. Masson, Paris, 499 pp
Millot G (1970) Geology of clays. Springer, Berlin, 429 pp
Millot G (1989) Proceedings of the IX international conference, Strasbourg
Mitchell D, Vincent A (1997) Exploration and appraisal of plastic sedimentary clays for the fine ceramics industry. Appl Clay Sci 11:311–327
Moll B (2001) Industrial clay mineralogy. CMS Workshop. Loyola University, Chicago
Moore DM (1996) Comment on: definition of clay and clay mineral, joint report of the AIPEA nomenclature and CMS nomenclature committees. Clay Clay Miner 44:710–712
Moore D, Reynolds R (1997) X-ray diffraction and the identification and analysis of clay minerals, 2nd edn. Oxford University Press, Oxford
Murray HH, Keller WD (1993) Kaolins, kaolins, kaolins. In: Kaolin genesis and utilization, Special publication No 1. The Clay Minerals Society, Boulder, Colorado, pp 1–24
Nickel EH (1995) The definition of a mineral. Can Mineral 33(3):689–690
Nickel EH (1996) Mineral names applied to synthetic substances. Eur J Mineral 8:461–462
Paquet H, Clauer N (1997) Soils and sediments: mineralogy and geochemistry. In: Paquet H, Clauer N (eds) Springer, Berlin, 369 pp
Patterson SH, Murray HH (1975) Clays. In: Lefond SJ (ed) Industrial minerals and rocks, 4th edn. American Institute of Mining, Metallurgical and Petroleum Engineers, New York, pp 519–585
Povarennykh AS (1972) Cristallochemical classification of minerals. Plenum Press, New York
Ross CS, Kerr PF (1931) The clay minerals and their identity. J Sediment Petrol 1:55–65
Sanfeliu TA (2002) Evaluation of the particulate atmospheric aerosol in the urban area of Castellon, Spain, Proceedings of the 5th international symposium on the protection and conservation of monuments in the Mediterranean basin, Sevilla, Spain. In: Galán E, Zezza AA (eds) Protection and conservation of the cultural heritage of the Mediterranean cities. Balkema, Lisse, pp 621–665
Souza Santos P (1975) Tecnologia de Argilas, v 1-Fundamentos, In: Blücher E (ed) Universidade dc São Paulo, Brasil, 340 pp
Velde B, Druc IC (1999) Archeological ceramic materials: origin and utilization. Springer, Berlin, 299 pp
Weaver CE (1989) Clays, muds, and shales. Elsevier, Amsterdam, 819 pp
Wilson MJ (1987) Handbook of determinative methods in clay mineralogy. Blackie, Glasgow, 308 pp
Worrall WE (1982) Ceramic raw materials, second ed., Institute of ceramics textbook series. Pergamon Press, Oxford, 111 pp

Further Readings

Gomes C, Silva J (2006) Minerals and human health: benefits and risks/Os minerais e a saúde humana: benefícios e riscos, bilingual edition of the authors. Multiponto, Porto, 142 pp
Velho J, Gomes C, Romariz C (1998) Minerais industriais: geologia, propriedades, tratamentos, aplicações, especificações, produções emercados. In: Velho J, Gomes C, Romariz C (eds) 591 pp

Chapter 3
Historical Aspects of a Natural Pharmacopeia: Clay in the Corpus of *The Medieval Pharmacopeia* Written in Arabic

Abstract This chapter opens with an historical presentation of some translations of Ancient Greek medical texts. They reached us partly thanks to translations into Arabic carried out during the Middle Ages. These texts were progressively enriched and thereafter evolved autonomously. Stories of conquests and religions, both ancient and medieval, are superimposed on the evolution of this knowledge, without being the exclusive causes. Therapies based on clay and its various daily uses are described in the corpus of profane texts of pharmacy, pharmacopeia, medicine, and rules for healthy living. In this chapter these points, disclosed in chronological order and by topics, are dealt with, concentrated only on some fundamental studies in medicine and pharmacy.

3.1 Introduction

The encountering of civilizations, Greek, Persian, Indian, and of many other cultures and thoughts of the Near East, Africa, and Andalusia, made it possible to establish a corpus of texts we here designate medieval pharmacy, written in Arabic.[1] The theoretical background of therapy during that time was based on the medical pathology of four humors systematized by Galen (131–216). For an historiography of Arabic pharmacy, see Hamarneh (1973) and Katouzian-Safadi (2004).

The political, territorial, and social conquests of the seventh and eighth centuries made it possible to elaborate a new Islamic and Arabic civilization, all the while knowing that the actors and protagonists spoke Arabic and other languages and that they practiced various religions. The linguistic instrument of these conquests, the Arabic language, rapidly led to the syncretism of sciences of this period (fusion of one or several cultures and religions into a common shared religion and culture), by opening new domains of application and by facilitating the diffusion and exchanges between peoples of various ethnic groups, religions, and languages.

Ever since the eighth century, works originating from remote geographical areas were gathered and translated into Arabic. The main independent sources of Indian

[1] There are two points here: (a) the dates are given according to the Gregorian calendar; (b) for the transliteration of Arab words, common nouns, or proper names, a simplified system is adopted to avoid specific Arabic diacritic letters.

M. Rautureau et al., *Clays and Health*, DOI 10.1007/978-3-319-42884-0_3

medicine and the *Book on Poisons* of pseudo-Shânaq translated into Persian and Arabic were widely dispersed. Thus, Indian drugs were mentioned in texts related to medical topics. The major works of Galen[2] (died around 216 CE) and the Treaty of Dioscorides[3] *De Materia Medica* (first century CE) were translated into Arabic. Moreover, the list of these medical products has been enriched over the centuries by at least 400 additional "drugs".

With the expansion of the geographical territory by Arabic and Islamic conquerors, from the Indus to Andalusia, new types of clays were incorporated into medical therapy resulting from local practices and medical empiricism.

The most important sacred text of Islam, the *Koran*,[4] frequently mentions soil (Toelle, 1999)[5] and clay (tîn ans tîna).[6] The soil when contrasted with the sky is very frequently designated *ard*[7] and used more than 450 times. But when it is contrasted with aqueous space, the term *barr* is employed (flora and humans emerged from *barr*). Clay is specified by various terms[8]: *tîn, salsâl, hamâ, turâb, sidjdjîl.*

Clay has various characteristics: it can be fine or sticky depending on the context. During the creation of man, malleable clay (or mud), *hamâ' in masnûnin*, was transformed into *salsâl* (solid clay); matter then evolves, just as in the potter's workshop. The powdery consistency of mud (*turâb*) becomes a denser and sticky clay (*tîn lazib*), then a malleable mud (*hamâ masnûn*), to be finally transformed without being fired, as *salsâl*, a pottery that emits a sound (*salsala*) when it is sharply knocked. Thus, during creation, a noble material was created from a more humble material such as *turâb* (dust), coagulated blood, and so on. Another material, *sidjdjil*, similar to clay or marl, is used to produce stones or *hadjar*.

[2] Hunayn Ibn Ishâq (804–877) is one of the most important translators of his time. He developed a scientific method of translation and gathered a group of translators around him; he wrote a missive (letter) that proves that major works of Galen in medicine and pharmacology were translated into Arabic by the ninth century.

[3] The treaty of Dioscorides, *Materia Medica*, translated by Stephan Ibn Basil, re-examined by Hunayn Ibn Ishâq, was translated from Greek into Syriac and Arabic. This treaty comprises 800 drugs coming from the three kingdoms, plant, animal, and mineral. Andalusian scientists played a major role in enriching this work. Ibn Djuldjul (died after 994) added 62 new drugs, mainly of Indian origin. An example is the myrobolans or camphor, odorous and disinfecting products. Later, Ibn al-Baytâr (died in 1248) enriched this list by 400 new drugs that then reached 1400 drugs. These new drugs (200 of plant origin and 200 of animal and mineral origins) were unknown to the Greeks.

[4] *Note*: For the Arabic terms, simplified transliteration is used as described in Appendix H. In our research, we have referred to the bilingual edition of the *Koran, Essai d'interprétation du Coran inimitable*, French translation by D. Masson, Gallimard, ed. 1980. We thank Khadidja Guellil for having indicated the salsal term to us, and for his revision of the Koranic terms.

[5] Toelle H (1999) *Le Coran revisité: le feu, l'eau, l'air et la terre*, Damas, Institut français de Damas.

[6] See *Encyclopédie de l'Islam* vol. 10, articles "Tîn,"and "Tîna," p. 568. For example see the term "Tîn" in the Koran, translated by D. Masson (*op. cit.*), verses 7, 12; 17, 38; 61, 76.

[7] Toelle H, "Terre" (or Soil) in *Dictionnaire du Coran*, Amir-Moezzi MY, ed. (2007), Robert Laffont, Paris, 867–869.

[8] Porter Y "Argile" (or clay), in Amir-Moezzi MY, ed., *op. cit*, 89–90.

With this vast range of terms, the Koranic text dealing with the history of religions and the rules of life does not mention any medical treatment making use of this product. This concurs with the vision of Ibn Khaldûn[9] (1332–1406) concerning medicine in the Islamic and Arabic civilization. In his famous book, *Muqaddima*, he describes the entire medieval society and in Chapter XXIV, under the heading of "Medicine," he exposes his extremely relevant vision. Ibn Khaldûn distinguishes the existence of various categories of medical practices: erudite medicine that associates theory, practice, and empiricism; empirical medicine without theoretical base; and medicine accepted as efficient [but not checked] thanks to our beliefs and convictions.

Given the precision and relevance of this passage, an extract is presented here:

> **Medicine**—It is an art dealing with the study of the human body from the point of view of health and disease. The doctor endeavours to preserve health and to look after diseases using remedies and food, after having identified the diseases related to each part of the body, the causes generating them and the remedies appropriate for each of them. As for the remedies, he guides himself on their composition and their strength, for the diseases, on the signs that indicate their seriousness, and for the patient on his aptitude to accept or refuse the remedies; this he does by examining the aspect of the patient, his excretions and his pulse…
>
> Galen is one of the greatest authorities in medicine of ancient times. His works were translated into Arabic.
>
> At the time of Islam, eminent physicians reached the top of this art, such as ar-Râzî, al-Majûsî and Ibn Sînâ. There were also many eminent doctors in Andalusia, among them the famous Ibn Zuhr. … Medicine belongs to the arts that are required only by urban civilizations and by luxury. … Men belonging to rural civilizations generally follow medical practices based on the limited empirical experiences of a few individuals. … Such medicines can include valid elements but they rely neither on natural rules, nor on conformity with the patient's temperament.

Thus Ibn Khaldûn places the activity of medieval physicians as a continuation of their predecessors, such as Galen, and quotes some of the major representatives of this science, to the East, Râzî (865–925) and Avicenna (980–1037), and to the West, Ibn Zuhr (1091–1161). To summarize, Ibn Khaldûn recalls an erudite medical history in which teaching and oral transmission are marked by the elaboration of great scientific and profane works. This "Art" is linked to urban civilizations and to all the tools available.

Therapies based on clay and its various daily uses are described in the corpus of profane texts of pharmacy, pharmacopeia, medicine, and rules for healthy living. With progress in differential diagnosis[10] and the discovery of "new diseases," clay became associated with new types of treatments. We examine these points in

[9] Ibn Khaldûn (2002). "Medicine," in *Le livre des exemples*, translated, presented, and annotated by A. Cheddadi, Gallimard, Paris, coll. Library of the *Pleiade*, 966–968.

[10] The construction and the expansion of large cities (such as Baghdad, Basra, Shirâz, etc.), the existence and construction of large hospitals (in Baghdad, in Ray close to Teheran), and the training of eminent physicians contributed very early to the definition and characterization of new diseases. The scientist Râzî, also known as Rhazès (865–925), is quoted for his critical judgment of diseases, their evolution, and their differential diagnosis.

chronological order and by topic. We concentrate here only on some fundamental studies in medicine and pharmacy. A complete historical overview of the Middle Ages on "clay" would require several essays on its own; such a study should be undertaken.

3.2 Geographical Varieties and Physical Properties of Various Clays

Empirically, medieval scientists differentiated various therapeutic qualities of clays depending on their nature. This was achieved by indicating the origins of clays, and—given the absence of modern technologies in physics—their physical characteristics were described: color, touch, smell, and reaction to water.

In his essay on pharmacopeia, *Al-Saydala fî al-tibb* (*Pharmacy in Medicine*),[11] the great scientist al-Bîrûnî (973–1050) who lived east of the Iranian plateau (present-day Afghanistan) listed eight categories of clay (*tîn*, in Arabic). A few centuries later, the botanist and Andalusian pharmacist of the thirteenth century, Ibn al-Baytâr (1197–1248),[12] listed nine categories of clays with very distinct therapeutic qualities. Among them, some had been known since the time of Dioscorides (first century CE) whereas others were discovered during the time of the Islamic and Arabic empire, after the seventh century.

Ibn al-Baytâr characterized clays based on where they came from, in order to facilitate their identification required for their use in medicine. He distinguished the following sources:

1. Island of Lemnos[13] (*tîn makhtûm* or terra sigillata)
2. Island of Sâmos[14] (*Shamûs tîn*)
3. Island of Chios[15] (*tîn djazîra al-Mustaka*)
4. From Egypt (*tîn Masr*)
5. From Sijilmassa[16]
6. From Spain

---

[11] Bîrûnî Abû al-Rayhân Muhammad b. Ahmad, *Kitâb al-saydala fî al-tibb. Al-Bîrûnî Book on Pharmacy and Materia medica*, Ed. with English translation by Said-Karachi HM, 1973; Ed. with comments of Abbâs Zaryâb. Teheran, 1981 (publication of the University of Teheran, No 572, Section History of Sciences, 3), article 'Tîn,' 411–415.

[12] Ibn al-Baytâr: *Djâmi' al-mufradât al-adwiya wa al-aghdhiya, re-ed. Beyrouth, 1492–1992, Traité des simples*, French translation of L. Leclerc, Beirut, 1412/1992, reprinted Institute of the Arab world, articles of "Tîn," vol. 2: 421–427, No. 1488–1496.

[13] Lemnos or Límnos is a Greek island in the northeast Aegean Sea.

[14] Sâmos is a Greek island in the Aegean Sea, near Asia Minor; today it belongs to Greece.

[15] Chios or Chio: Greek island in the Aegean Sea, near Turkey.

[16] Sijilmassa is a city located in the oasis of Todgha in Morocco.

7. From Kimolos[17] (*tîn qîmûlîyâ*)
8. From the town of Seleucia[18] in Syria
9. From Armenia close to Cappadocia (*tîn armanî*)
10. From Neyshâbûr (or Nîshâbûr, or Nîshapûr), in the northeast of present-day Iran
11. Soil or clay of vines (*tîn kermy*), not related to geography but to cultivation of the vine

Reading these texts encourages us to consider certain names and sources with caution. The "clay of Kimolos" must be considered as a generic name, indicating certain apparent properties of purity and certain therapeutic effects. This clay is frequently called *hurr* by the inhabitants of Basra. Moreover, the "clay of Kimolos" can come from the island of Kimolos (a Greek island) but other areas could produce clay with the same properties. Thus the name of *qîmûlîyâ* or "Kimolos" could be attributed to material coming from Armenia or Sijilmassa (a city located in the oasis of Todgha in Morocco) or from the Iberian Peninsula.[19]

Clays coming from the East were usually not employed in the West. In the thirteenth century Ibn al-H'achcha[20] edited a dictionary explaining the medical terms used in *Al-Mansûrî,* a medical encyclopedia written by the eminent scientist Râzî[21] who lived in Eastern Arabia (area of present-day Iran and Iraq). Ibn al-H'achcha designated the following five types of clay:

1. *Mulabban* clay (or milky clay): It is consumable, but not very much appreciated.
2. Clay qualified as *hurr*: It is clay sifted to remove sand and stones.
3. *Makhtûm* clay (or terra sigillata): It is known in the East but not in the West.
4. *Akil* clay (or clay to be consumed): It is clay to be orally consumed; there are multiple kinds of clays with the same name, but they have similar qualities; none is famous in the West where only children consume it.
5. Clay of Khurâsân (clay from an area located in northeast present-day Iran); It belongs to the consumable types, unknown in the West.

[17] In Arabic *Tîn qîmûlîyâ*: chimolea. Island of Kimolos or Cimolos (in: *Islands of Greece*, Lacroix L (1853), Original from the New York Public Library, digitalized, p. 472) and in *Précis de géographie universelle* (1835), Paris, ed. Aimé André, of Malte Brun, p 135, one can read: "Cimolos, re-elected for its clay: the soil cimolienne."

[18] Seleucia is the name of many cities that were founded during the Hellenistic period by the sovereigns of the dynasty of Seleucid, in particular in Syria, Mesopotamia, and Anatolia.

[19] See article "Tîn kîmuliyâ" (Thîn Kîmolya, Terre Cimolée or cimolée soil), Ed. Leclerc L, 2: 423–424.

[20] Ibn al-H'Achcha (thirteenth century BC), *Glossaire sur le Mansûrî de Razès*, Ed. with introduction by G. S. Colin and H. P. J. Renaud, Rabat, Imprimerie Economique, 1941, see "Tîn" (Clay), p 6.

[21] Râzî, Muhammad ibn Zakariyyâ B. Yahya al-, *al-Mansûrî fî al-tibb*, with ed., and comments by al-Siddîqî HB, *Manshûrât ma' had al-makhtûtât al-'arabiyya*, Kuwait, 1987.

It should be noted that according to this author, three types of clays frequently employed in the East are rarely used in the West. Thus clays are approached by the author Ibn al-H'achcha with precaution depending on their origins, and such an approach demonstrates a high degree of accuracy in observation and practice.

A famous institution known as *Hisba*[22] was in charge of monitoring and verifying the administration of cities: bridges, irrigation canals, roads, food, and so on. It also had the responsibility of pharmacy supervision and of ensuring the quality of drugs, particularly of expensive products that were imported or manufactured on the spot. It had to fight against falsification. The *muhtasib* (inspector) used many chemical techniques to test the purity of the products. Abu al-Munâ, living in the thirteenth century in Egypt, and known as Cohen al-Attâr, was one such pharmacist-inspector, the author of a book written in 1260 that was intended to help detect falsifications of drugs.[23] It provides precise details on clays, for example:

1. Among clays,[24] terra sigillata (*tîn makhtûm*) is recognized by its alum smell; in his book Cohen al-Attâr states that it has a red to yellow color.[25]
2. Soil coming from Egypt (*tîn Masr*) must be rich in ashes.
3. Clay from Kimolos (*tîn qîmûliyâ*) is white; it has a fatty aspect and feels cold when touched; it is compact and difficult to mix with water. If it is black, it is not appropriate for therapeutic use.
4. Soil from Neyshâbûr is white and does not become sticky in the mouth; it has a good salty taste that disappears when in contact with fire.

3.3 Which Clays for Which Therapies?

According to the theory of the (four elements built up gradually over several centuries in Egypt and in the ancient Greek civilization), the sublunary world is composed of four elements common to both inert and living matter. These four elements—fire, air, earth, and water—are characterized by their specific qualities: hot, dry, cold, and moist.

---

[22] See "Hisba," in *Encyclopédie de l'Islam*, ed. Brill Maisonneuve & Larose (1971), 3: 503–510.

[23] Some examples are: the stone lapis lazuli should not change color and aspect when burned, its powder should not color water; balsam oil is very expensive (it is worth twice its weight in silver): when deposited on a cloth, it does not spread but remains in the shape of a bubble (probably thanks to its surface tension).

[24] Abû al-Munâ Dawûd Ibn Abî Nasr al-ma' rûf bi Attâr al-Harûnî, *Minhadj al-dukkân wa dastûr al-a'yân fi a'mâl wa tarâkîb*, Ed., Hassan 'Âsi dâr al-manâhil, Beirut, Lebanon, (1412H/1992) 320 pp; see for clay, 'Tîn,' 246–247.

[25] Medieval authors give several indications: geographical sources, consistency when touched, odor, and color. We know that the designations of clays according to their color (white, pink, green, etc.) currently do not correspond to any particular structure or composition.

The humoral theories,[26] known to the Egyptians, were clearly recognized by the Greeks from the fourth century BC. According to the latter theory, circulation of the four liquids (designated humors), phlegm, blood, yellow bile, and black bile, occur in the human body. In each body, there exists a correspondence between humor and the quality of the four elements; their imbalance causes disease.

Matter in general and medical substances in particular are composed of the four elements. Their resulting qualities depend on their composition. The substances can act on the balance of humors, and consequently on both health and disease states depending on qualities in excess. Each substance can reach four degrees of quality, the fourth degree being as strong as poison. Today, this theoretical framework corresponds neither to our scientific knowledge (in chemistry and physics) of matter, nor to human physiology. However, on the basis of precise observations of patients and of the empirical results obtained, the physicians of ancient times and of the Middle Ages managed to prevent diseases and to cure patients. At a given time, a theoretical framework "very distant" from the one that dominates our current scientific knowledge, could lead to relevant empirical observations,[27] and to efficient treatment of pathologies (Grmek 1980, Gourevich and Grmek1985).

Thus, clay can be used alone, but very often it is part of a composition in which it is associated with other drugs. As a simple product, clays are known to be moderately dry and cold[28] (in the second degree). In what follows, texts of certain famous physicians illustrate their use of clay therapeutics. It is important to emphasize that in no case did the physicians draw any conclusions on a general use of clays or subscribe to their permanent use.

3.3.1 *According to Hunayn Ibn Ishâq (804–877)*

A famous translator of Galen's texts, Hunayn Ibn Ishâq[29] produced several compilations[30] as well as new texts. One of his original texts very frequently studied by physicians is concerned with ophthalmology.[31] In this text, Hunayn gathered the

---

[26] See humors, 'Humeurs' in D. Lecourt (Ed.), *Dictionnaire de la pensée médicale*, Presses Universitaires de France, 2004, 598–603.

[27] Grmek, MD (1980) "A plea to free the history of scientific discoveries from myth." in *On Scientific Discovery*, 9–42, M. D. Grmek et al., Publishing Company, Boston, and D. Gourevitch and M. D. Grmek, (1985) "Les expériences pharmacologiques dans l'antiquité," *Archives Internationales d'Histoire des Sciences,* 35, 3–27.

[28] There exist four qualities (hot, cold, dry, and moist). Each quality has four degrees as described by Galen.

[29] *op. cit.*

[30] See, for example, *Les questions sur la médecine*, a book of propedeutics for physicians, by Hunayn Ibn Ishâq, *al-masâ' it fî al-tibb*, Muhammad Ali Abû Rîyân, Mercî Muhammad Arab, Djalâl Muhammad Musa, Ed. Dâr al-djâmi'ât al-misrîyya, Beirut, Lebanon, 1978.

[31] Hunayn Ibn Ishâq, *Kitâb al-'ashara al-maqâlat fî al-Ayn al-mansûb li-Hunayn Ibn-Ishâq, The book of the ten treatises on the eye ascribed to Hunain Ibn Ishâq* (809–877 AD), ed. in Arabic with translation into English by M. Meyerhof, Cairo, 1928, p. 222 Arabic, p 227, English.

remarks of Greek physicians, especially of Galen, and enriched them with some personal experiences. Chapter X of the book is devoted to compound drugs used to treat the eye; clay is a component of these ointment.

Clay of Sâmos[32] is incorporated in a bandage for the eye to cool, reinforce and drain the eye and to, increase the density of the oinment. Various creams are proposed to treat ophthalmia, various keratitis with hypopyon, cornea ruptures, or iris prolapses. One of them is attributed to the seventh century physician Paul or Bulos (in Arabic), and is composed of cadmium, white lead, rose petals, Arabic gum, "tragacanth,"[33] aloe, and clay from Sâmos.[34] It should be stressed that Hunayn generally used clays employed by Greek authors. On the other hand, the use of clay from the Iranian plateau, annexed after the seventh century to the zone under Islamic and Arabic influence, does not appear in these texts.

3.3.2 According to Râzî (865–925)

Râzî was born in Ray, a town close to present-day Teheran. He was Persian, wrote in Arabic, and was given the name "Galen of Arabia" because of his knowledge of medicine, his diagnoses, and his therapies including all the conditions of the patients (environmental as well as psychological).

Medieval bibliographers mention a text of Râzî specifically written on clay of Neyshabûr, a town of the region of Khurâsân in eastern Iran. Sections of this lost study entitled "The benefit of clay in tablets"[35] are quoted by other physicians. Thus Ibn Samadjûn (end of the tenth century) recalls that this clay was prepared in the form of tablets having sometimes the shape of birds; it was perfumed with camphor or musk, and was eaten when drinking wine (as an appetizer); it gave a pleasant smell and calmed stomach pains. It was known to decrease vomiting and to calm difficulty in digestion after sweet and fatty meals. Its consumption is more beneficial if it is accompanied by ingredients appeasing the stomach and releasing a pleasant smell such as that of rose petals, common galingale, or sedges (*Cyperus longus*),

[32] See the book by Hunayn Ibn Ishâq, *Kitâb al-'ashara maqâlat..., The book of the ten treatises on the eye..., op. cit.*, p 198 Arabic, p 131 English, "*al-tîn al-mansû 'ilâ Sâmûs al-muqallab bi al-kawkab*" (clay of Samûs considered as star-shaped clay).

[33] Adragante is a deformation of the word tragacanth. The gum obtained from plants of the genus *Astracantha* (or *Astragalus*) in *Kitâb al-ashara maqâlat..., The book of the ten treatises on the eye..., op. cit.*, p 211 Arabic, p 143 English.

[34] Keratitis with hypopyon is a form of infectious keratitis; there is often loss of substance of the type of serpiginous ulceration of the cornea and presence of pus inside the front anterior chamber.

[35] Râzî, Arabic title of the book: *fî an li al-tîn al-mutinaqil bihi manafi'*; no manuscript of this work is to be found. For this passage we refer to the article in Persian of M. Mohaghegh, entitled "Râzî va al-tîn al-Neychabûrî" (or Râzî and the clay of Nichabur), *in*: *Collected Texts and Articles on the History of Medicine and Medical Ethics in Islam and Iran*, Soroush Press, Teheran, 1995, 337–348.

snap ring (*Cymbopogon schoenanthus*), cubeb (*piper cubeba*), or cardamom (*Elettaria cardamomum*). Compared to other clays, it has the advantage of not producing renal lithiases (renal stones); however, people with renal weakness, or a fragile liver, that is, those who often appear thin and have a tanned, slightly yellow or green skin should not consume this clay because it would diminish their appetite and their skin would become more yellow.

The historic book testifies that some people liked to consume this clay in great amounts and this gave them a yellow color and certain disturbances. This famous clay constituted one of the prides of the population of Neyshabûr in the Middle Ages.[36]

One of Râzî's works, addressed to medical students, is *al-Mansûrî* that was translated into Latin several times and taught in the universities of Paris and Montpellier during the Middle Ages. In this book, Râzî often referred to clay, and insisted that it should be carefully administered. He knew the dangers of intestinal obstructions if administration was inadequate. Thus he specified that edible clay could create occlusions of hepatic channels and serious indigestion,[37] whereas the same clay strengthened the cardia (upper opening of the stomach).[38] Terra sigillata and clay from Armenia were regarded as efficient against any abnormal rejection of blood by hemoptysis[39] or melena.[40] The author quoted clay as medicine with drying capacity, but neither aggressive nor corrosive.[41] Based on the work of Galen, Râzî[42] recommended this product accompanied by wine in the case of pestilentia.[43] For purges, this physician proposed clays from Sâmos, Faritus, and Tûs.[44]

In other major works, clays were included in compound drugs. The art of composing drugs was considered difficult; physicians therefore explained these rules in a specific book or in a chapter of their books related to medicine. One of their theoretical problems was the pathways the product should follow within the body to reach the organ concerned; this implies passing all the barriers of digestion. It was necessary to accompany the medicine (healing product) with another one acting as a vector.[45] Clay is often used as a vector to transport an active product to the organ

---

[36] Mohaghegh M, *op. cit.*, p 341.

[37] Râzî, *Al-Mansûrî op. cit.,* p 174.

[38] The cardia is the anatomical term for the part of the stomach attached to the esophagus.

[39] Hemoptysis: Rejection of blood resulting from a cough, for example.

[40] Melena: Evacuation of blood from the anus.

[41] Râzî, *Al-Mansûrî, op. cit.*, p 193.

[42] Râzî, *Al-Mansûrî, op. cit.*, p 227.

[43] Pestilentia: generic term for all infectious diseases accompanied by high fevers, causing epidemics and deaths. This term could at times indicate the particular case of plague. See M. D. Grmek (1983) *Les maladies à l'aube de la civilisation occidentale*, Ed. Payot, Paris: 29–31.

[44] Tûs, a city located in the region of Khurâssân, a province in the East of Iran; the clay of this region was not mentioned in the Greek medical texts.

[45] Avicenna describes this function of transporting a medicine in his book *Qânûn* (*Canon*); for example, saffron binds to camphor, helps it reach the heart, and then departs. See *al-Qânûn*, (*op. cit*) 3, p. 309.

suffering from hemorrhage. In the case of lung or throat treatments, clay should be accompanied by aromatic herbs.[46]

Râzî was the first to propose a differential diagnosis of two diseases, measles and smallpox.[47] His book, written in Arabic, was translated into Greek,[48] Latin, and various European languages.[49] In his original book, *Measles and Smallpox* (in Arabic, *al-hasba wa al-djudarî*), all the stages, etiology, symptomatology, and the patient's psychology, as well as the majority of the treatments were invented from the discoveries made by Râzî; clay was one of the drugs used in health care. Thus, in the event of smallpox, it required that the doctor pay special attention to the joints where the bone, ligaments, and tendons were in danger. He proposed an ointment containing clay from Armenia.[50] In Chapter 13, a compound drug is administered[51] to reduce diarrheas and prevent patient fragility. One of its important ingredients is clay from Armenia.

Râzî used his general experience of medicinal therapy to treat the symptoms of measles and smallpox. Clay was often used in these therapies.

---

[46] Al-Sulamî quotes Rhazes on the rules of medicine composition. Al-Sulamî, Abd al-Aziz, *Question and Answers for Physicians, a Medieval Arabic Study Manual*, translated, Ed. with introduction by G. Leiseir and N. R. al-Khaledy, Brill, Leiden, Boston, 2004, Arabic p 51, English p. 79.

[47] Râzî, *Kitâb al-hasba wa al-djudarî*, or *The book of measles and smallpox*, translation into Persian by Mahmûd Nadjmâbâdî, University of Teheran, 1040, 3rd ed., 1992.

[48] Congourdou MH, "The Greek translator of the treaty of Rhazès on smallpox," in *Storia e ecdotica dei testi medici greci*, Atti del II Conv. Intern. (Paris, 1994), A. Garzya (Ed.) Napoli: M. D'Auria ed., 1996. 465 pp 99–111.

[49] In the manuscript of the BNF, Paris 2228, the Greek translation attributed to Joannes Actuarius (end of thirteenth century) seems to have been ordered by an emperor of Constantinople. In Europe, this book was first known thanks to its Greek translation, which was then translated into Latin. A Latin edition was produced in 1556 in Venice with the title *De pestillentia*. A first Latin translation from Arabic was made by Mead in 1747. Other translations follow: that of Hunt (the original manuscript is unknown), that of Iohannis Channing, *De variolis et moribilis* with the Arabic text for the first time (1766), the translation of Greenhill, *A Treatise of the Smallpox and Measles, Rhazès* (1848), and finally, the French translation of J. Leclerc and A. Lenoir, *De la variole et de la rougeole* (1866).

[50] Râzî, *The book of measles and smallpox* (*op. cit.*), Arabic p 18, line 24, and p 19, line 1. This ointment contains the following ingredients: sandalwood, Glaucium (in the family Papaveraceae), clay of Armenia, rose petal, camphor, vinegar, and rosewater. It must be strictly applied to the articulations only.

[51] Râzî, *The book of measles and smallpox* (*op. cit.*), Arabic p. 27, line 5. This drug contains: crushed rose petals, bamboo, sumac (*Rhus Coriaria*), spine barberry, Arabic gum, clay of Armenia and terra sigillata, bark of poppy, and pomegranate; it is accompanied by concentrated acidic quince. change OK

3.3.3 *According to Avicenna (980–1037)*

Avicenna (Latin for the Arabic Ibn-Sînâ) was a philosopher and physician who became famous through his book on medicine, *Qânûn* (*Canon*).[52] In this work he described 12 types of clays[53] in the chapter dealing with simple drugs.

For clays, qualified as "edible" or *tîn ma'kûl*, Avicenna confirmed the remarks made by his predecessor Râzî.[54] This clay obstructed the digestive tract, and Avicenna proposed avoiding it because its frequent consumption could even create a harmful dependence for the patient.

The clay of "Crete," *tîn aqrîtash*, regarded as less efficient than the other clays, "decreases mental capacity." It is efficient for pimples, redness, and itching of the eyelids.

Qîmûliya clay can be white or purple. It loses its corrosive character once washed. Mixed with vinegar, it is beneficial for burns, wounds (provided it is used before the formation of abscesses), or blisters. It is beneficial for articulations and tumefactions at the level of the joints.

Avicenna often mentioned terra sigillata as *antipoisonous*, referring back to Galen, and proposed using this clay against dog bites and various other bites. Authors preceding or succeeding Avicenna mentioned the antipoison quality of this clay with the same rhetoric, such as Abû al-Munâ[55] when he writes about *tîn makhtûm* (or terra sigillata):

> It can be found in a cave in an island of Byzantium. Before its use, the turbid part of its water suspension is removed by decantation and it is clarified; mixed with goat blood,[56] it is dried and stamped with a seal bearing the name of the king of the island. It is used in theriaca [a compound drug]. Galen specifies that when he crossed this island he did not observe the kneading of clay with goat blood. He therefore questioned people on this subject without obtaining an answer. On the other hand, he saw that after washing, the clay is dried; a paste is made from this clay and water, and it is then transformed into tablets. It is stamped, i.e. certified, in the same locations as stamping for kings. There is a difference between this clay and the other clays: the use of other clays stops haemorrhages (ishâl) while this one facilitates evacuation by eliminating the poison.[57] As with the others it is astringent. I saw it only once, it was red turning to yellow and it carried the seal.

---

[52] *Qânûn* (*Canon*) (*op cit.*) rapidly became a book to teach medicine in the East as well as in the West, in particular in the universities of Montpellier and Paris; see D. Jacquart, "La réception du Canon d' Avicenne: comparison entre Montpellier et Paris aux thirteenth and fourteenth siècles," *in: Actes du 110e congrès national des societés savantes*, Section of history of sciences, 1985, Paris, II: 69–77.

[53] *Qânûn* (*Canon*), *op cit.* I: 328. The first chapter of the second speech of this book is reserved for the current pharmacopeia. These simple drugs are classified according to a certain order of the alphabet (*abjad*). In his introduction, the author states that he describes current and well-known drugs and that he does not innovate. This is why this chapter is interesting: it gives a complete view of the drugs and experience of physicians before Avicenna.

[54] *Qânûn* (*Canon*), *op. cit.*, p. 329.

[55] Abu al-Munâ, *Minhâdj al-dukkân*, p 247, *op. cit.*; this is based on our translation.

[56] The addition of goat blood to clay is mentioned in many texts, but for the moment we cannot explain this addition. According to this text, Galen raised the same question.

[57] Because of its capacity of evacuation, clay is added to the "theriac," a famous electuary against poisons.

3.3.4 From the Eleventh Century Onwards

The Jewish Andalusian Ibn Biklarîsh, of the eleventh to twelfth centuries, wrote a book on simple drugs.[58] He enumerated five categories of clays and described how to prepare one of them:

> Certain soils also are stony and they should be washed until they are pure and the stones totally removed. They still need to be more thoroughly washed when they are to be eaten or prior to their use for rectal injection than before they are used to treat tumours. Thorough washing is as follows: take a sample of the soil containing stones, add water, mix the fine soil with water and stir the mixture strongly and continuously, and finally let it settle. Once the deposit is formed, remove the water, take out the soft and viscous matter, abandon the stony and sandy components concentrated at the bottom, and dry the tender and viscous matter.

This example of the washing and preparation of clay demonstrates the classical and very efficient current method of separation of the various size components by "decantation."

The tradition of referring to clay in books on pharmacopeia continues in modern and contemporary times. Thus in a work published in Persian at the end of the eighteenth century by Aqîlî Khorâsânî,[59] under the heading *Clay* there is a general description followed by 20 types of clays defined by their geographical origin, their physical characteristics, their therapeutic qualities, and their various applications.

3.4 Clay and Daily Life

3.4.1 Omnipresence of Clay

The principal person in charge of the institution known as *Hisba*,[60] the security authority in charge of the administration of medieval cities, had to send inspectors to all the markets of the city to supervise the various trade associations. Clay was one of the products the inspector had to check because it was commercialized by various merchants.

Weavers made special baskets to transport clay, known as *tîniyyât*.[61] Craftsmen used moulds of clay to produce duplicates from original keys. Clay vases were recommended for the conservation of rare and expensive electuaries (compound drugs

---

[58] Ricordel J, *Al-Musta'ni: Les tables des médicaments simples of Ishâk Ibn Biklârish* (letters alif à zây), ed., translation and analysis. PhD thesis in epistemology and history of sciences, University Paris 7, Denis Diderot, 25 November, 2004; for article *"Tîn"* or clay; see the manuscripts folios [72r–73v].

[59] Aqîlî Khurâsânî, MH, *Makhzan al-adwiyya*, or Pharmacopeia, Calcutta impression, 1844, offset 1375/1993, Teheran: 588–595.

[60] See *Hisba* (*op. cit.*).

[61] Al-Sakatî de Malaga, *Un manual Hispanique de Hisba,* Arabic text with notes and glossary, ed. Colin GS and Levi-Provençal E, Paris, 1931, p 64, line 16.

for drinking) for several years. Râzî (who was also a chemist) proposed a particular preparation of clay to improve the distilling equipment making it more hermetic:[62] first it was mixed with triturated excrements, animal hair, and specific amounts of various salts; then it was fermented in a moist atmosphere to obtain a new product with a better sealing capacity.[63]

The food products were heavily supervised in the markets. Certain meals were consumed on the spot and others were "to be taken away." The container used to roast meat could not be of lead or copper. For roasted meat intended to be sold, the inspector had to verify that the interiors of the furnaces were properly coated with clay of a satisfactory quality and kneaded with clean water, so as to guarantee the quality of the cooked meat.[64]

Avicenna, in his *Qânûn* (*Canon*),[65] quoted uses of clay other than therapeutic. The soil of Armenia was commonly used by goldsmiths to dye gold. In agriculture, painting vines with "vine clay" or *tîn al-karm* protected the leaves and buds from attack by worms.

3.4.2 Clay: Often Present in Fakes

Craftsmen manufacturing laced boots, sandals, or slippers could cheat by placing clay between two pieces of animal skin to thicken the material artificially;[66] it was difficult for the market inspector to detect this counterfeit.

The use of red clay to falsify various therapeutic products was known to the inspectors of pharmacies. False ointment was produced by certain ophthalmologists by mixing colored products with red clay.[67] Dragon's blood (a red gum composed of several species of plants),[68] was an extremely expensive product. One way to produce false dragon's blood was to heat liquefied resin[69] with powdered fired brick and red clay.

---

[62] Garcia-Sanshez E, "Les techniques de distillation de l'eau de rose à al-Andalus" in *Res orientales XI, Parfums d'Orient,* Peeters ed, Bures-sur-Yvette, 1988:125–140.

[63] Al-Râzî, *al-Madkhal al-ta'lîmî*, ed. English transl. and comment by Stapelon HE, Azo RF and Hedayat Hoseyn M, *Mem. of the As. Soc. of Bengal*, VIII, 1927, p 414.

[64] Shayzarî, al-'abd al-Rahman, *The Book of the Islamic Market Inspector*: *Nihâyat al-rutba*, translated with an introduction and notes by Buckley RP, Oxford University Press, 1999, p 54.

[65] Avicenna, *Qânûn* (*Canon*; *op. cit.*).

[66] Al-Sakatî de Malaga, *Un manuel Hispanique de Hisba* (*op. cit.*) p 64, line 16.

[67] Shayzarî, al-'abd al-Rahman, 1999, *op. cit.*, p 118.

[68] Dragon blood is a red gum that comes from the island of Socotra in Yemen; it is provided by *Calamus draco*, of the family of palm trees; Ibn al-Baytâr, ed. Leclerc, *op. cit.*, vol. II, p 94, No. 882.

[69] It seems to be the resin of Coccus or *Carteria Laca*, Shayzarî, al-'abd al-Rahman, 1999, *op. cit.*, p 68 notes 31.

3.4.3 Clay: Ingredient of Legendary Cures

As a curing agent, clay is legendary as also are honey and wine, and allowed more than one physician to become wealthy.

One can quote, for example, the legend of Bakhtîshû' Ibn Jibrâ' îl, the physician of the caliph al-Mutawakkil (847–861). When the caliph complained about stomach burns, the physician put clay, water, and food in a calabash and slightly warmed the entire mixture before it was consumed. Food then posed no more problems. Clay provided this physician with three hundred thousand dirhams!

N.B.: This chapter was voluntarily limited to the corpus of the medieval texts translated from Greek to Arabic and to the original writings in Arabic and Persian; this in no way excludes knowledge and know-how from all the traditional civilizations that are dealt with in many essays. Our aim is to show that a parallel can be drawn between elaborate contemporary scientific knowledge, and older empirical knowledge. This contemporary scientific knowledge can provide some explanations. At times empiricism was built on a base that seems inaccurate today; it was then necessary to note it and leave it to the scientific disciplines to try to look for new explanations. In life sciences and particularly in the care and treatment of human bodies, ancient knowledge should be treated with the greatest respect and relevance. One can then observe that civilizations that had no contact with each other came to use the same clays: the roots of this knowledge are based on a great capacity of observation during healing. Therefore, these treatments deserve the attention of contemporary scientists. This is what can be described as the integration of tradition in science.

References

Hamarneh S (1973) Origins of pharmacy and therapy in the Near East, Tokyo. The NAITO Foundation, English Publications, Series 1, p 2

Katouzian-Safadi M (2004) Pharmacie arabe. In: Lecourt D (ed) Dictionnaire de la pensée médicale. Presses Universitaires de France, Paris, pp 870–874

Chapter 4
Description of Clay Materials

Abstract Clay origin and nomenclature and clay modes and conditions of formation are discussed in this chapter. Also the typology of clay deposits and the quality of clay, in terms of purity, required for clay uses in the field of human health are dealt with as well. In addition to the almost general beneficial effects of clay and clay minerals it is acknowledged that some clay minerals could have real or potentially harmful effects on the respiratory system if inhaled, due to the fibrous character of their particle shapes. The study of clay and clay minerals' quality requires a thorough knowledge of both principles and analytical methods (chemical, thermal, textural, microscopic, diffractometric, and spectrometric) used in mineralogy in order to identify and characterize the crystallochemical specificities of clay minerals, the crystallochemical formula of those natural clay mineral species utilized in the field of health exhibited in the present chapter. Clay exploitation also takes into account clay utilizations in various domains such as pedologic, crafts, and industrial and chemical, all depending upon clay mineral crystallochemistry.

4.1 Definition of Clay

Through the ages clays have been received definitions adapted to the state of knowledge at the time and the uses to which they were put. As universal materials they cannot be limited to only one definition. The vocabulary of every language is rich in characteristic and descriptive terms; all these minerals are common to all our activities. Close to the word "clay," a generic universal name is that of "earth," associated with qualifying names such as potter's clay, fire clay, or green clay. The term "clay" comes from the Greek word *argilla*, the root of which, *argos*, signifies "white," the color of kaolin clay used in ceramics. A relation to purity was quickly established with this color even though most clays are not white.

The attribution of a name to a clay material still remains an actual subject. A nomenclature of systematic character exists corresponding to a classification developed by Caillère and Hénin (1963) and accepted by the majority of scientific committees who sometimes adapted it to specific technical and professional needs. Nevertheless, the same clay often occurs in different geologically similar places. Consequently, how can we avoid the attribution of the name of the place where the deposit occurs or of the person who discovered or studied the clay deposit?

M. Rautureau et al., *Clays and Health*, DOI 10.1007/978-3-319-42884-0_4

Moreover, within the same deposit, clay can have a variable composition depending simultaneously upon the number and extension of the defects specifically related to the substitution of the structural ions occurring in the clay minerals, the fundamental constituents of clays.

For more precise details about this subject the authors of this monograph advise those interested to read the works of Caillère and Hénin (1963); Caillère et al. (1982); and Millot (1964), as well as the recent treatise, *Handbook of Clay Science* (2006), Bergaya F, Theng B, Lagaly G (Editors).

It would be preferable here to use the clay denomination based on the crystallochemical characteristics of the principal clay mineral present in the clay. Unfortunately it is not easy for the general public and, consequently, the current names are sometimes useful because they are more easily recognized and remain associated with these "more erudite" definitions.

The real state of clays, currently well known, is essentially described on a microscopic scale. For this reason clays remained very difficult to study and understand until the beginning of the twentieth century.

The concept of clay, when considered as a material constituted of microparticles, should emphasize the properties related to variable size scales according to the type of study:

1. Intrinsic properties of the crystalline structure and the chemical composition of an elementary layer, and limited to an ideal unit cell (dimension of 1 or 2 nm)
2. Properties calling upon the stacking of elementary layers in the form of single-crystal microparticles (dimensions of some micrometers)
3. Properties implying an important mass of material, and the macroscopic whole of a very great number of particles assembled in more or less compact aggregates

In the actual stage of our knowledge, it is still difficult to establish a theory of unified modeling that allows easy passage between the properties suitable for each of these scales. It seems that this difficulty will last for a long time because the complexity, quite real, has several origins:

1. Of a physicochemical nature
2. On the other hand of a theoretical and mathematical nature
3. And related to the extreme variability of the real situations where these minerals intervene

Nevertheless, sharing old empirical results and those already acquired by modern physicochemical sciences provides increasingly satisfactory achievements, and allows, in many cases, justifying knowledge that hitherto was empirical.

In parallel to the description of clays, more exactly of the hydrous phyllosilicates, many finely divided minerals have, in certain aspects, similar behaviors. Actually they are behaviors resulting from the characters imposed by the crystalline structure or by the morphology (essentially the size) of the particles. Certain properties such as thixotropy[1] or rheology[2] are not specific to clays, but they often justified the initial studies.

The presence of clays is closely associated with the geology of the surface layers of the Earth. Their genesis and their constant evolution make them a source of raw materials whose reserves are large and varied. This does not mean that they are inexhaustible in the field of the specificity of the varieties or their quality. Clay's capacity of swelling in the presence of water acts in an important way on the soil mechanics' stability. Regarding the use of clays in the field of health, homogeneous deposits, basically untouched and relatively important, are necessary to ensure the safety of therapeutic actions. Due to their microdivided state it is advisable to be careful with the deposits because the treatments for clay purification are expensive and not always possible. It is this uncertainty of provisioning that justifies the recent preference for the synthesis of certain varieties, despite their high commercial prices.

4.2 Geological Context: Origin of Clays and Formation of Clay Deposits

Most clays have a geological origin; for this reason they are natural materials. Users' requirements are increasingly precise on the definition and purity of clay minerals. Recent research led to the synthesis of some of these minerals; it paves the way in the future for particularly precise applications. However, the price of artificial clays is certainly in favor of natural clays and will be for a long time.

Clay minerals' components are not spontaneously and necessarily present in all rocks. They are found close to the Earth's surface where they develop primarily at the time of weathering and hydrolysis of depth-formed silicate rocks. Consequently, they are initially distributed in dispersed form in very small volumes of rocks (surface outcrops, fractures, or cracks) subjected to natural constraints, in particular with the action of water (hydrothermal origin or rainwater). After their formation at the expense of the pre-existing altered rocks, they generally undergo transport via the surface waters that gradually sweep them along. They accumulate finally in more important sedimentary deposits, nearer or farther away from the formation or source sites, at the bottom of lakes, seas, or oceans. The sedimentary layers are often very thick. In the course of time and increasing burying, they are compacted by their own weight and the matter mass that covers them little by little. In accordance with the theory of the general geological cycle, today firmly established, and well-described by Millot (1979), some layers will be able to undergo various reworkings and sometimes rising or burial with movement of the tectonic plates. Hence, new rocks are unceasingly formed, that will later be at the origin of new clays. The

[1] Thixotropy is the property related to the formation of gel.

[2] Rheology is the study of the plasticity of highly hydrated clay minerals.

natural conditions of the cycle formation–transport–accumulation–burial–uprising lead either to a mixture of minerals of various origins, or less frequently to very pure deposits.

The existence of clay deposits is almost universal in the zones where water is (or was) present. Note that the large silicate geological solid masses have allowed the formation of thick sedimentary deposits,[3] where several mineral species are side by side, some of them clay minerals. An analysis by X-ray diffraction and an observation by electronic microscopy allow the identification of the species that are simultaneously present. The use of these two techniques provides invaluable information for eventual uses and for the purifications necessary before any marketing.

The composition and stability of clays are in direct relationship to their formation conditions. It is thus very important to know these factors if one wishes to explain the reactions of clay placed in a situation that is not usual for it (eating, e.g.). From this point of view it is necessary to require precise mineralogical information, even structural, from the industrialists who provide the clays meant for medical or paramedical use.

Without going into more precise detail, we must approach the description of the various modes of clay formation. We retain three of them: heritage, transformation, and neoformation.

- ***Heritage***: When clays are exposed to rigorous weathering on their place of formation, winds or the run-off waters move and transport part of them to the sea or to lacustrine zones. The modifications are limited during this transport, thus these clays are known as inherited clays. They have properties in connection with the nature of the weathered bedrock and the climate of the continental areas of formation.
- ***Transformation (degradation and aggradation)***: After transport and accumulation in discharge points (detrital deposits), during a certain period of time the clay particles can undergo transformations that induce a change of their mineralogical state, and vary according to the context. Thus, an acid medium can destabilize the initial structure of clay minerals and leave only one residual siliceous skeleton. The eliminated octahedral cations could participate in the formation of other composites. In other cases, it is the fixation of ions that prevails. For example, a smectite or vermiculite that could fix an alkaline ion would reconstruct mica, whereas the fixation of hydroxyl ions on the same smectite involves its transformation into chlorite (the structure of this clay mineral is shown in Fig. 5.7).

  Transformed clays are frequent in the soils where the weathering fronts (containing clay) located between nonweathered rock and weathered rock migrate downward as the soil thickens at the expense of the parent rock. Some transformations, natural but more complex, can result from bioreactions. These complex reactions have more or less expected effects. Pedology shows such evolutions in soils where clays and organic matter are side by side and constantly interact. In the subject that concerns us here and during clay's use in connection with a living

[3] Impregnated thickness of a field in geology is the parameter that expresses how thick a geological or mineral deposit is.

organism, such transformations must not be neglected. In the event of ingestion, for example, intestinal microorganisms can cause these reactions, and thus modify the type of mineral sometimes quickly, and simultaneously its properties.
- ***Neoformation***: In nature, when an ionic solution that is present in the porosity of the rocks reaches favorable concentration, pressure, and temperature conditions, clay minerals can precipitate directly from the saturated solution:[4] this is called neogenesis. The stability of the formed clay products depends strongly on the environmental conditions at the time of this genesis. Therefore for health care it is useful to know the origin and mode of formation of clays to judge their nature and their properties. However, these conditions are not met in the physiological media that are the subject of the present report.

4.3 Quality of Clay Deposits and the Harmfulness of Certain Clay Minerals

Clay applications are quite numerous in the field of health but, in tonnage, they represent a very modest quantity compared to that estimated for all clay industrial applications.

N.B.: To avoid confusion between technically useful clay minerals, recognized as such, with mineral species known as harmful or likely to be so, we voluntarily use the name "phyllosilicate" in the case of the latter.

Sometimes there exists a certain ambiguity on the beneficial or harmful effects of clays, with the image of all the therapeutic actions that act on living organisms in pain. Actions in the field of one therapy may be detrimental for another. Thus, certain types of clay have textures[5] that can appear dangerous, such as the aforementioned case of chrysotile whose fiber action on the pulmonary or digestive membranes leads to the appearance of mesothelioma. To reach such severe pathologies implies that the action of the phyllosilicate was strongly concentrated over a long period of time, as was the case for the workers who handled significant amounts of asbestos under badly controlled conditions. We do not necessarily need to worry people who have simply come near an object made of asbestos or had contained some, but it is necessary to remain careful during such handling. On the other hand it is wiser to eradicate these sources of contamination and modify the technologies employing these materials. The important matter is that of the destination of the products extracted during decontamination of sites containing asbestos. Dumping chrysotile in specialized ground is only one very insufficient temporary solution. The change of technology and the eradication of old wastes are neither always simple nor feasible and can lead to true socioeconomic dilemmas. In spite of everything, the usefulness of these materials remains indisputable in many technological applications, even knowing that they are dangerous for health.

[4] Solution contained in the porosity of the rock.

[5] "Texture" refers to the shape of elementary clay particles: sheets, fibers, and the like.

As a precautionary measure, and for the same reasons, the effects of other fibrous phyllosilicates with pseudo-layers such as sepiolite and palygorskite must be attentively supervised in their applications in relation to the organism.

By analogy with well-known diseases such as silicosis, prudence is also essential for the physical status of certain clays, and it is then advisable to extend the precautionary principles. If clays are used in the dry dust state (in powdery form, easily dispersed in the air), it is dangerous to allow them pulmonary access. In all this handling, the wearing of a mask is recommended. In fact, if silica is considered as resulting from the attack of clay by an acid, clay particles of small sizes can be comparable to the microparticles blamed in silicosis. In the same way, the specific chemical bonds of the clay's surface can generate overly strong affinities with the membranes of the organisms they are brought next to, and they then are able to upset the local balance. This is why any attempt at medication always results in putting in parallel the beneficial effects and the doubtful or the harmful ones that result from too repeated use. Only physicians are qualified to make these decisions; if necessary they will require information from clay scientists and will draw the relevant conclusions.

Our major focus in this book is health and the means of preserving it in the best possible state. It is thus necessary, on these criteria, to point out very clearly the danger in utilizing some phyllosilicates (at least in certain types of uses).

The clays and clay minerals utilized for health purposes must present a good degree of purity and be able to be supplied systematically. This does not always happen and many products are not sufficiently controlled. They often remain in a rough state that can generate serious difficulties at the time of use. It is necessary to say in fairness to the producers that the industrial demand for high quality raw materials is very strong. This makes it difficult to reserve certain deposits whose purity is recognized for use in the field of health. It is not easy or may be impossible to set up regulations in this field but it could become necessary. Scientific knowledge of the properties should make it possible to reach that point.

4.4 Mineralogical Characterization

Although studied in detail for a long time by geologists, clay minerals remained rarely studied by mineralogists and crystallographers. The Laboratory of Mineralogy of the National Natural History Museum of Paris contributed a great deal to their study, in particular due to Professor Lacroix.

In fact, the mineralogy of clays was dealt with in a satisfactory way only from the years 1930 onwards, thanks to the development of the following.

1. Chemical analyses and research on the chemical bond (Caillère 1933; Pauling 1930, 1949)
2. Thermal analyses (Mackenzie 1957)
3. X-ray diffraction analyses (between 1930 and 1950, roughly)
4. Spectrometry analyses, using various techniques, particularly infrared

The reliable classification of clays that resulted from these analyses was presented by Caillère and Hénin in 1959, and since then they have not undergone major modifications.

For many years the essential character of clay description has been the size of their constituent particles: in fact, the clay particles in general have an average size less than 2 μm, the limit of the optical microscope's resolution. This definition, necessary but too simple, has led to assigning to clays other materials of the same dimensional class but of completely different composition, nature, and properties. A common extraneous mineral in the sediments is finely divided quartz. Technically, it is possible to separate the coarse particles mechanically but a very fine fraction always exists.[6] Certain minerals can be eliminated by chemical treatment as in the case of magnesium or calcium carbonates, aluminum, and iron oxy-hydroxides when they are subjected to a diluted acid attack.

Figures 4.1 to 4.4 illustrate the characteristics of the internal and external organization of microparticles of kaolinite, the principal clay mineral constituent of kaolin, characteristics that were obtained through four types of scientific data. These techniques of mineralogical determination of clays, resulting from physics, are considered routine today. The power of the conclusions that they initially could bring was essential; it was used by mineralogists for the determination of the various species and for

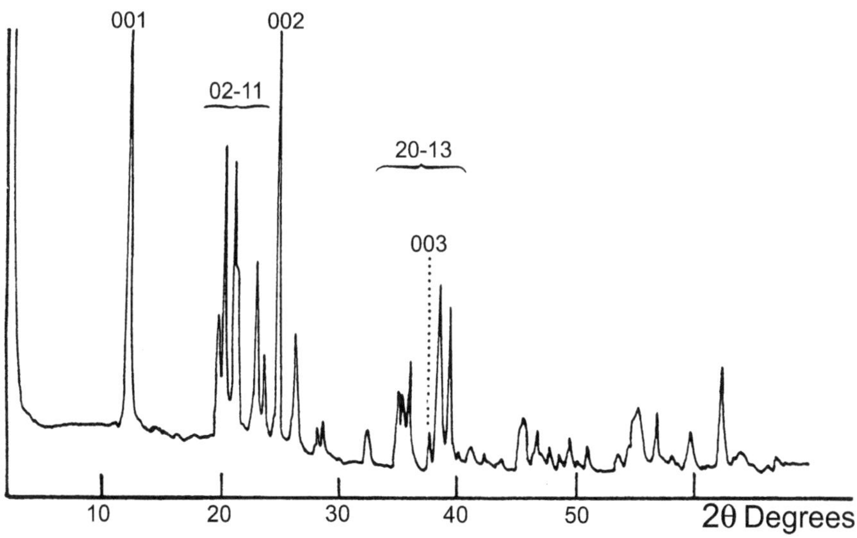

Fig. 4.1 X-ray diffraction pattern of a structurally well-ordered kaolinite

[6]Quartz exists in the form of very small particles smaller than one micrometer and often nanometer size. Remember that the angstrom (whose symbol Å) is a unit of length equal to 10^{-10} m, or one ten millionth of a millimeter, very close to the magnitude of the distance between two neighboring atoms in a nongaseous chemical compound. However, the legal unit in this area is the nanometer which we strive to use although traditionally clay specialists use the angstrom.

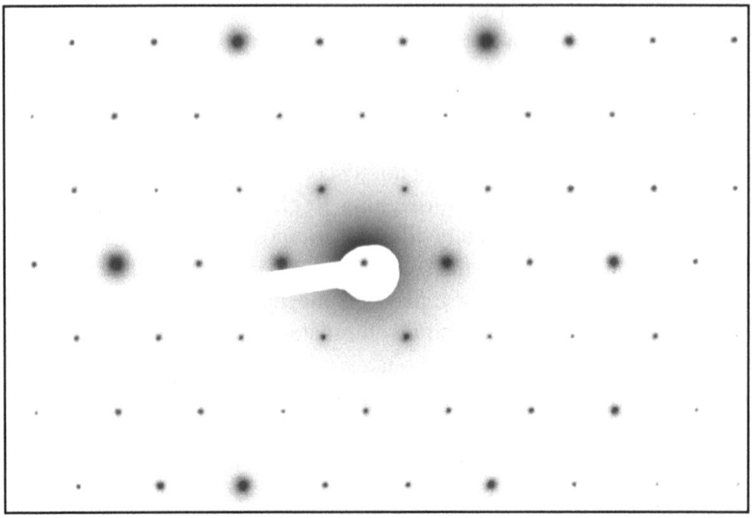

Fig. 4.2 Stereotype electron microdiffraction pattern of a structurally well-ordered kaolinite

Fig. 4.3 Kaolinite microparticles observed by transmission electron microscopy (TEM)

physicists to provide powerful bases of reasoning on the properties of clays. Clarifying them for unspecialized users remains an open problem.

Methods initially using X-rays made it possible to highlight the crystalline character of clay minerals. This fundamental character is illustrated in Fig. 4.1 by an X-ray diffraction diagram obtained on a kaolin powder (rock). The peaks or

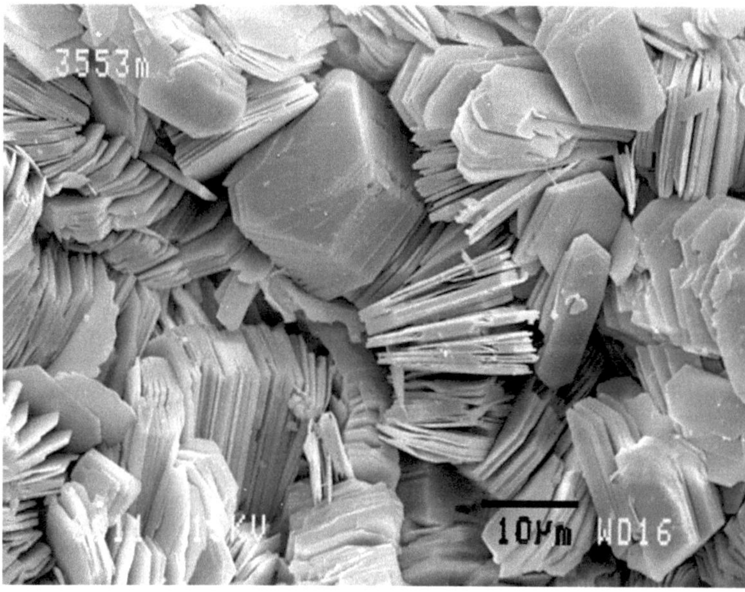

Fig. 4.4 Kaolinite particles observed by scanning electron microscopy (SEM)

diffraction maxima unambiguously indicate the crystalline nature of a clay mineral: structurally very well-crystallized kaolinite. More recently electron microscopy made it possible to obtain solid confirmation of this character thanks to the possibility of simultaneously reaching the microdiffraction of electrons on single-crystal individual microparticles (Fig. 4.2) and the visualization of the microparticles (Fig. 4.3). Lastly, the regrouping of layers in the form of particles and the association of these particles in aggregates are illustrated in Fig. 4.4 which was obtained using scanning electron microscopy. All these characters are undeniable; they opened the way to a scientific exploitation of the properties of clay minerals. These characters offer important scientific prospects in clay utilization in the field of human health; they are well known to modern specialists but they are not easy to grasp for the general public.

Clays are thus truly minerals and they can be classified. However, the study methods needed to become sufficiently accurate to adapt the determination techniques to their very small dimensions. Scientists had to move towards multiscale studies of the properties of the material. This concept of relationships between properties and particle dimensions is rather recent and, to exploit the physical and chemical properties of clays, it is advisable always to bear in mind the scale of the studied phenomena.

Without calling into question previous knowledge, these new approaches open a scientific possibility of studying the properties of clays, particularly in the field of health. We hope that this projection is a solid support to choose and develop the use of clays in the medical or paramedical field.

Structural and chemical formulas[7] of principal minerals used for health care are given below (for more details, consult Appendix D):

1. Kaolinite $(Si_4)^{IV}(Al_4)^{VI}O_{10}(OH)_8$
2. Smectites:

   Montmorillonite $\left(Si_8\right)^{IV}\left[Al_{4-y}Mg_y\right]^{VI}O_{20}\left(OH\right)_4, M_{y+}$ octahedral substitutions

   Beidellite $\left(Si_{8-x}Al_x\right)^{IV}\left[Al_4\right]^{IV}O_{20}\left(OH\right)_4, M_{x+}$ tetrahedral substitutions
3. Talc $(Si_8)^{IV}(Mg_6)^{VI}O_{20}(OH)_4$
4. Illites $\left[Si_{(8-x)}Al_x\right]^{IV}O_{20}\left(Al_4\right)^{VI}\left(OH\right)_4\left(K\right)_x$ tetrahedral substitutions
5. Sepiolite $\left(Si_{12}\right)^{IV}\left[Mg_8\right]^{VI}O_{30}\left(OH\right)_4\left(OHH^+\right)_4 8\left(OH_2\right)$
6. Palygorskites $\left(SiAl\right)_8^{IV}\left(MgAlFe^{3+}\right)_5^{VI}O_{20}\left(OH\right)_2\left(OHH^+\right)_4, 4\left(OH_2\right)$

N.B.: Sepiolite and palygorskite have structures with pseudo-layers; their formulas are given for ½ unit cell.

They are the ideal formulas relative to theoretical crystallographic models for pure clay minerals constituting what are called di- or trioctahedral "poles". In the majority of natural clay minerals, there are substitutions that are shown in the developed formulas described below.

For example, the crystallochemical formula of the model and that of a real sample analyzed in the laboratory are given below for the TeOc family of kaolinite. These writings give a more documented approach than those presented in general papers or on packaging. We clarify this case:

Ideal crystallochemical model:[8] $(Si_4)^{IV}(Al_4)^{VI}O_{10}(OH)_8$

Exploitation of this formula leads to the following ponderal contents:

$$46.55\ \%\ \text{of}\ SiO_2 \quad 39.49\ \%\ \text{of}\ Al_2O_3 \quad 13.96\ \%\ \text{for}\ H_2O$$

The formula can also be written in another form known as the "equivalent oxide formula," which represents the quantity of oxides likely to be analyzed:

$$4SiO_2\ 2Al_2O_3\ 4H_2O$$

This formulation does not reflect the ordered state of the crystalline matter (and sometimes results in wrongly regarding clays as sources of silica or alumina which is partially false).

[7] The notation $(Si^{4+})^{IV}$ means that the Si^{4+} ion has a coordination number 4 and is at the center of a tetrahedron. In the same way $(Al^{3+})^{VI}$ means that the Al^{3+} ion has a coordination number 6 and is at the center of an octahedron; M is a compensation ion.

[8] The Roman superscript numbers represent the state of coordination of the elements; the others refer to their valence.

In reality, there exist substitutions and analyses obtained, for example, for the well-known kaolinite[9] of Saint-Austell (Cornwall), such as the following,

$$46.2\ \%\ \text{of}\ SiO_2 \quad 39.2\ \%\ \text{of}\ Al_2O_3 \quad 13.8\ \%\ \text{for}\ H_2O$$
$$\text{with traces of}: TiO_2(0.09\ \%);\ Fe_2O_3(0.23\ \%);\ MgO(0.07\ \%);\ CaO(0.06\ \%);$$
$$Na_2O(0.09\ \%)\ \text{and}\ K_2O(0.21\ \%).$$

This leads to writing the formula detailed in the form:[10]

$$(Si_{3.98}Al_{0.019})^{IV}(Al_{3.96}Ti_{0.006}Fe^{3+}{}_{0.015}Mg_{0.009})^{VI}O_{20}(OH)_4, Ca_{0.006}Na_{0.015}K_{0.023}$$

This formulation shows the complexity of the problem of the structural attribution of the chemical elements. However, a simplification appeared gradually through a finer and more and more specific analysis. Thus the presence of calcium, sodium, and potassium could be allotted to traces of coexistent mica beside kaolinite. Titanium oxides such as anatase and rutile (TiO_2), but also oxides of other metals such as hematite (Fe_2O_3) are, as a rule, omnipresent in the exploited deposits, in the form of micro- or nanoparticles. Hence, Ti and Fe in very low amounts could integrate the structure of kaolinite. To consolidate clay minerals' chemical formula, a very fine mineralogical analysis is thus necessary within the framework of pharmaceutical industrial usage. In addition, very thorough analyses show that there is a segregation of the various other minor elements,[11] and that certain zones of the layers collect titanium, for example, whereas magnesium is concentrated in other zones.

Such formulas must be based on studies affirming that the elements taken into account really belong to the structure of the clay mineral or that they come from "polluting" grains. When these elements are present independently of the clay mineral, in the form of mineral impurities, their properties are clean and are thus potentially usable in other respects.

When moving from the microscopic field to the macroscopic field (that which we see with the naked eye), the material is perceived on different scales. If the material is dry, then it is compact and hard but is nevertheless composed of very fine grains gathering a great number of small particles such as those visible in Figs. 4.3 and 4.4. From Fig. 4.4 one can then imagine a very vast clay mass characterized by connected porosity[12] between parallel stackings of elementary platy particles. This

[9] Kaolinite of Saint-Austell, Cornwall, England (analyses realized at the Laboratory of Mineralogy of the National Natural History Museum of Paris under the direction of Professor Simonne Caillère).

[10] This type of writing corresponds to a natural mineral (and not to a model) and makes it possible to reveal the existing substitutions highlighted by the chemical analysis. The reading of these formulas must be understood according to a statistical ratio of distribution on the sites because it is obvious that there cannot be fractions of atoms on a given position.

[11] Analyses by electronic microscopy (using the spectroscopy of the losses of energy).

[12] When the channels of the porous environment are joined to the others, it is said that they are connected. In this case, a fluid can move easily in this porosity.

porosity is accompanied by important capillarity properties. In the presence of a liquid, most generally water, a mechanism of capillary forces causes the aspiration of the liquid inside the clay mass. Water initially comes in contact with the packages of layers and, according to their nature, creates various types of connections. Initially the more easily accessible layer edges are hydrated and the interlayer ions are hydrated later on. In the case of kaolinite which does not inflate, and where water does not penetrate the interlayer space, absorption is limited only to the edges of the layers and the porous space located between the particles. When the interlayer space is penetrated by water, the properties of the particle change because water allows the packages of layers to overlap each other: it is the plasticity characteristic of wet clays. The clay mass becomes deformed and keeps its new form more or less for a long time according to the quantity of water it contains. It can even pass from a malleable medium to a thixotrope gel. Obviously these characters are important in geology and pedology. They are less so in the field of health. For many years they were the cause of difficulties in understanding the behavior of clays; we also find characters of plasticity and of thixotropy again in medical care applications.

Now we know better that these properties and the clay minerals typically seem to have multiscale properties. Each application is in relation to a dimensional scale that is clear for it. To make an error of assignment of scale at the time of use of these minerals can lead at best to failure, but sometimes to dangerous situations. Prudence is essential and requires the intervention of specialists in clay minerals and, if necessary, a probatory time of experimentation.

We must draw the attention of the reader to another essential characteristic of clay minerals: that of the existence of mixtures of layers of various types within the same particle. By convention these minerals are known as interstratified. Their existence reflects the natural changing conditions of genesis. The particles characteristic of an "interstratified" clay mineral do not represent a mixture of various clay minerals, but a succession of layers whose order of stacking is variable. In this form, it is impossible to separate the layers from the different varieties and we must take the mixture into account in its globality. This situation is current and during the collection of clay minerals on the ground, it is possible to find such minerals interstratified in the rocks. They are not immediately recognized and only complex analyses make it possible to identify them. We do not approach interstratified clays in detail; it is necessary to know they exist and represent an important stock for the future but that thorough studies are needed to use them in the field of health. This work is already being seriously advanced by specialized laboratories.

4.5 Exploitation of Clays

Faced with the multiple industrial employment of clays, the big question remains as to the quality and quantity of the supplies of raw material. For a long time geologists indexed the localization of the clay layers and studied their conditions of formation in detail (Millot 1964, 1979). The mineralogists and crystallographers then learned how to specify more and more finely the nature of clay minerals (group, family, variety). Later on crystallographers and geochemists were able to define the position of the ions and substitutions, making it possible today to study the physicochemical properties users hope to develop (Handbook of Clay Science, 2006, 2013). Users more often do not have access to correct information. Reading the specifications of the contents on a package of clays shows that improvements are possible with up-to-date knowledge.

To mitigate the uncertainties related to natural provisioning, some difficult experiments requiring know-how but already promising, were carried out in the field of clay synthesis. Compared to the therapeutic needs, they aimed at combining better control of the properties of clays and those of the surroundings or added molecules. The main difficulty was in controlling the formation conditions, but progress is being made. Economic conditions still lead to the use of "natural" products, always available in sufficient quantities and inexpensive to exploit; a disadvantage lies in their mineralogical variability resulting from the conditions of their genesis. Let us remember that although these minerals are natural they are not inexhaustible. Great attention is absolutely necessary so that their employment is most judiciously possible for the benefit of all.

4.5.1 Pedological Field

Clay minerals are essential components of soils; their physicochemical properties confer determining roles on them. Soils are media that contain many living microorganisms. The presence of clays is useful for their development thanks to their properties of fixation of water and ions. This remark constitutes a notable aspect in favor of the harmonious coexistence of living organisms and clays (or at least of the majority of them). The idea is indeed tempting within the framework of the exploitation of the presumably beneficial natural functions, but one must remain careful on this point.

1. First, clays present properties of adsorption allowing the fixation of ions or molecules possibly followed by exchange phenomena with ions or molecules from the environment. The reversible retention of water by clays is an essential and vital property for human activities, because it supports the development of plants on soils (Hénin et al. 1969).
2. Second, the electric balance of the crystalline structure of clay minerals is ensured by compensation ions fixed on the layers. These ions play a big role in pedology. Their mobility supports the development of plants in their function of consumption. In this process, many mechanisms are put in action so that the soil reaches a dynamic stability, such as the ionic exchanges by dialysis or diffusion through the membranes which come into play when living tissues and clays are put into contact.
3. Soil clays also constitute an oxidation agent of the humic components found there.

Many studies of tropical soils showed that an inappropriate exploitation of the soil involves an impoverishment of essential components and sets up a dangerous imbalance by increasing the relative concentration of unutilized components. Thus intensive agriculture causes the accumulation of chemical elements due to leaching or destruction of various minerals. In the health field, great geographical areas seem associated with the pathologies generated by excesses or deficiencies of chemical elements in the soil such as fluorine, iodine, arsenic, or nitrates (this list is by no means exhaustive). In the majority of the zones concerned, clays are affected by alternations of rainy and dry periods that wash and disaggregate the ground. A review of these phenomena was carried out by Dissanayake and Chandrajith (1999). These effects on human health are beginning to be better understood and analyzed; they still have to be taken into account and controlled if possible. In this context of the raisoning, clays play a very important part.

In the animal world, it should be noted that, in extreme or difficult situations, certain animals can find rock salt which is formed or settles on the walls of the caves, for example. They also know spontaneously how to find water by trampling the ground and thus causing water to swell by capillarity from ground containing fine particles and clays, as is shown with elephants. Termites are also experts in this search for biogenic salts associated with water.

The field of pedology is rich in lessons. The contact between the earth and living organisms is one of the most natural. There is still much to be learned and understood in this relation. The biomineral interface is a true physicochemical factory.

4.5.2 *Craft Industry and Industrial Field*

Clays constitute one of the very first raw materials used by craftspeople and industry. Natural or transformed they permit innumerable applications. We consider here only those that refer to health.

In general, the industrial field employs clays that are transformed by heating. These clays or derivative products are not directly related to health. However, very many medical and tableware utensils are part of our health concerns because they make it possible to obtain a general hygienic quality. Humanity is certainly much indebted to these utilitarian ceramic products containing clay that have been manufactured for many years. It is a reminder in the field of health and a very important aspect of the applications of transformed clays.

We can also indicate high-tech ceramic products initially resulting from phyllosilicates used in dental or osseous prostheses.

Without being in direct contact with the organisms, clays are used on a large scale in their natural state by the food industries, first of all, in both filtration and purification processes. The smoothness of their particles makes it possible to obtain a porous environment able to retain extremely fine particles. Thus, the first effective filtration cures, for the water provisions of big cities, were carried out on sand cushions and clays:[13] it was then a question of fighting against cholera-type

[13] Dehérain P (1892) Traité de chimie agricole. Ed. Masson, Paris 1, 640–650.

diseases. The effectiveness of this filtration results not only from low dimensions of the clay particles but also of the activity due to their walls. This technique is still in use today. For more controlled performances, a special porous ceramic based on clay makes it possible to constitute candle filters, which are very effective in providing drinking water, but in small quantity.

In certain food manufacturing industries, oil purification and discoloration are usually carried out using smectite-type clays. Talc, phyllosilicate without atomic substitutions, therefore neutral, is used to avoid adherence and in particular to produce "unsticky" rice. Perfectly stable and safe gels are produced in the manufacture of some foods (and also animal food).

4.5.3 Chemical Field

The modern chemical industry is an important user of clay on a twofold basis:

1. As a raw material to carry out mineral loads.
2. As a reactive agent. Clays are catalyst supports, even effective catalysts as well as ionic or molecular exchange supports.

These particularly interesting surface properties of clays justify the essence of their use by industrial or even fine chemistry. The oil industry and petrochemistry use huge quantities of clays for drilling and oil cracking.

As regards our subject, clay is part and parcel of a chemical reaction cycle where it acts as a reactive component. When it is found in digestive processes where it decreases the rate of acidity, it can also be dislocated itself if the medium is sufficiently acid. It acts, on two planes, by a buffer role to fight against stomach acidity (required effect) while the residues, resulting from clay after having been attacked by this acidity, can have a considerable secondary role in the case of repeated and abusive use (an associated effect of which one must be aware). It is the case of the colloidal silica that can then be formed. It is thus necessary to balance the beneficial effects against those of the generated ailments. Undoubtedly it is the too-frequent recourse to their use that can become harmful by accumulation.

In the manufacture of perfumes clay is sometimes included for its adsorption and fixation properties. On the same principle, unpleasant odors can be trapped, for example, when clay powder is placed inside a refrigerator.

References

Bergaya F, Theng BKG, Lagaly G (eds) (2006) Handbook of clay science, 1st edn. Elsevier, New York

Bergaya F, Theng BKG, Lagaly G (eds) (2013) Handbook of clay science, 2nd edn. New York, Elsevier

Caillère S (1933) Contribution à l'étude des minéraux des serpentines, Thèse de doctorat d'Etat *es* sciences. Muséum National d'Histoire Naturelle, Paris
Caillère S, Hénin S (1959) La classification des argiles, congrès géologique international, Mexico.
Caillère S, Hénin S (1963) Minéralogie des argiles, 1ère édition. Masson, Paris
Caillère S, Hénin S, Rautureau M (1982) Minéralogie des argiles, 2ème édition, 2 tomes. Masson, Paris
Dehérain P (1892) Traité de chemie agricole. Ed. Masson, Paris, pp 640–650
Dissanayake CB, Chandrajith R (1999) Medical geochemistry of tropical environments. Earth Sci Rev 47:219–258
Hénin S, Gras R, Monnier G (1969) Le profil cultural. Masson, Paris
Mackenzie RC (1957) The differential thermal investigation of clays. Mineralogical Society, London
Millot G (1964) Géologie des argiles. Masson, Paris
Millot G (1979) Les argiles. Pour la Science 20:61–73
Pauling L (1930) The structure of micas and related minerals. Proc Natl Acad Sci U S A 16:123–129
Pauling L (1949) Nature de la liaison chimique, traduction desmaroux. Presses Universitaires de France, Paris

Chapter 5
Clay Reactivity Depending Upon the Crystallochemical Properties of Clay Minerals

Abstract The crystallochemical models of clay mineral structures are described in this chapter because clay properties and inherent reactivity depend entirely on the specificities of those structures. Clay reactivity is strongly conditioned by the surface properties of clay mineral particles that are basically dependent on the particle global electric charge and its spatial distribution pattern, different from basal surfaces to edge surfaces. Particle electric charge results mainly from atomic substitutions that occur in both the octahedral and tetrahedral layers. In clay minerals bearing interlayer spaces chemical reactivity also possibly occurs on internal surfaces.

The reactivity of the clay–water system is quite common in natural clays. It is often used in mud therapy and peloid therapy in the form of paste for topical therapeutic and cosmetic applications, and it is also studied as well as the clay–water system that in the form of dispersion ("argillic water") is used for internal therapeutic purposes. Clay's ionic exchange capacity and color, two important properties whenever clay is used in the field of health, are also discussed.

5.1 Clay Mineral Structures

5.1.1 Ideal Crystalline Configuration of Clay Mineral Structure

Properties of clays can be approached and studied differently according to their use. These properties are very different from clay to clay, its composition and the scale of use.

The ionic crystalline structures of clay minerals are presented in the form of planes of atoms gathered in sheets that in turn build superimposed layers: phyllosilicates. This aspect is essential to understanding clay mineral properties that differ according to whether the basal or edge faces of their particles are considered. Finally, we especially note that clays are always microcrystallized minerals whose reactivity is primarily conditioned by the surface properties of the particles. The reactivity on a macroscopic scale, conditioned by the presence of water, generates typical thixotropic properties specific to gels, muds, or pastes.

M. Rautureau et al., *Clays and Health*, DOI 10.1007/978-3-319-42884-0_5

To understand the current applications of clays, or to find new ones, it is necessary to know and understand the crystalline structures of the clay minerals that compose them. Thus it is necessary to start by defining this concept of structure. In a general and very brief manner, two large families of compound materials can be identified:

1. The first corresponds to compounds having a known number of atoms connected to each other in a limited volume: these are molecules.
2. The second is that of the crystals where the position of the atoms or ions that compose them are repeated periodically according to an elementary unit cell; in theory one cannot define a limit in a crystal; it extends at infinitum. In reality, the size of the crystals is always largely dependent on the conditions of formation and their associated defects. Clay minerals, which we deal with here, are crystalline particles that generally occur under heterogeneous and variable natural conditions; thus they have many defects.

The presence of clay minerals in the crystallized world makes it possible to explain most of their properties. To begin, we present those elements essential to understanding this.

First it is a question of describing the ideal theoretical framework without using overly technical details but being precise enough to present the elementary bricks of the fundamental structure. The latter is articulated around two structural entities always present: tetrahedrons (Fig. 5.1) and octahedrons (Fig. 5.2).

More generally $\left[R^{2+\text{ or }3+}O_6\right]$

$$R^{3+} = Al^{3+}, Fe^{3+} \ldots \quad R^{2+} = Mg^{2+}, Fe^{2+}, Ni^{2+} \ldots$$

The tetrahedrons join to give a "tetrahedral layer" plane (Fig. 5.3) including in its thickness three planes of ions (respectively, O^{2-}, Si^{4+}, O^{2-}). This arrangement leads to a hexagonal plane symmetry characterized by the existence of a basal zone empty of ions called the "hexagonal cavity," which is fundamentally important in clay properties. In effect, this zone allows a molecule or an external ion that is near the surface to feel the anomalies of the electric field developed by the internal ions of the structure. For this part of the presentation, we retain only the silicon (Si^{4+}) and oxygen (O^{2-}) ions. It is obvious that electric balance is not acquired under these conditions but our objective concern for the moment is only the essential entities that constitute the layer. Their formulations are discussed after this description.

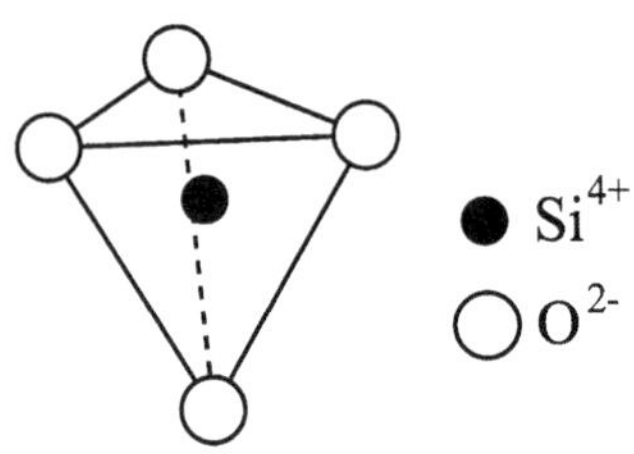

Fig. 5.1 Schematic representation of a tetrahedron $[SiO_4]$

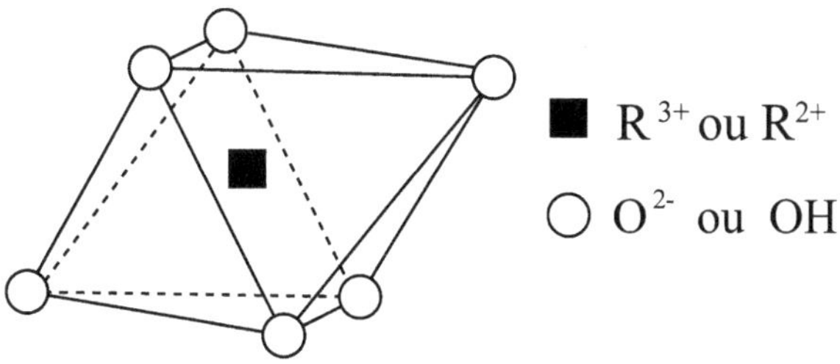

Fig. 5.2 Schematic representation of an octahedron [$R^{n+}O_6$]

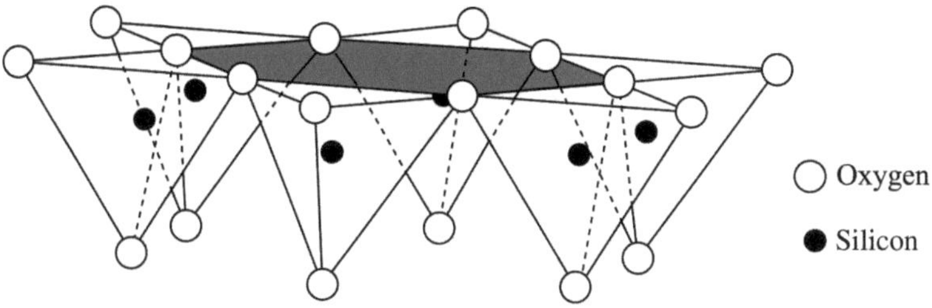

Fig. 5.3 Diagram representing the tetrahedral sheet and showing the hexagonal hole (represented in *gray*)

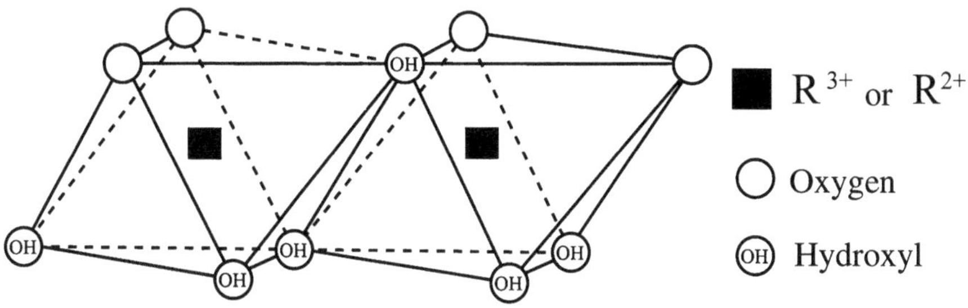

Fig. 5.4 Diagram representing the octahedral sheet

On the same principle, the octahedrons join to form an "octahedral layer" (Fig. 5.4) that also contains three planes of ions (O^{2-} or OH^-, Al^{3+} or Mg^{2+}, O^{2-} or OH^-) in its thickness.

$$R^{3+} = Al^{3+}, Fe^{3+} \ldots \quad R^{2+} = Mg^{2+}, Fe^{2+}, Fe^{2+}, Ni^{2+} \ldots$$

These two structural units (tetrahedral layers and octahedral layers) cannot remain in this state; they join together to form phyllosilicates that are more stable on the energy level.

By gathering a layer of tetrahedrons and a layer of octahedrons, we obtain a first unit that is the basic crystalline structure of the clay minerals of the group to which kaolinite belongs (Fig. 5.5). In brief, it is a mineral whose elementary layer is of the

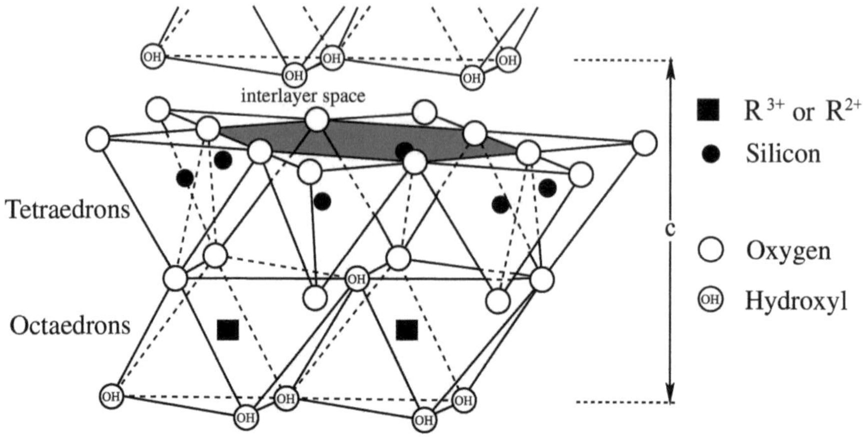

Fig. 5.5 General diagram representing a TeOc-type clay mineral layer

type named TeOc (a layer of tetrahedrons and a layer of octahedrons). This type of structure has a particular characteristic: the two higher and lower faces of the same layer are completely different. One is constituted of oxygen whereas the other is formed only by hydroxyls. The hydrogen ions present in the hydroxyls make this unit electrically neutral. The hydroxyls located between the tetrahedral and octahedral layers, placed in the middle of the layer and subjected to strong connections, are thus not very mobile. The hydroxyls located on the external octahedral basal surface of the layer are more easily mobile and immediately evoke the mobile presence of water.

$$R^{3+} = Al^{3+}, Fe^{3+} \ldots \quad R^{2+} = Mg^{2+}, Fe^{2+}, Ni^{2+} \ldots$$
$$\text{Kaolinite when } R^{3+} \text{ is aluminum}$$

A contrario, we do not find this diversification for the TeOcTe-type phyllosilicate layers for which the two basal faces are equivalent (Fig. 5.6).

The representation adopted for Figs. 5.6 and 5.7 clearly shows the possibility of a plane extension of the mineral and symmetries associated with it. One also observes the planes of atoms there (here in their ionic form) and their organization in the form of a sheet (of tetrahedrons and octahedrons). According to the case, the gathering of two or three layers constitutes an elementary layer, and the standard perpendicular superposition of the same type of layers allows the development of particles in the third dimension. In the drawings we have shown distance "*c*" between two layers: it is the fundamental distinctive lattice parameter of the various families of phyllosilicates (called the "interlayer specing or interlayer parameter").

When the TeOc layer is supplemented by a new tetrahedral layer on the other side of the octahedral layer, we obtain a TeOcTe mineral whose standard model is that of the illites (small grain size micas), of the smectites (various ions in

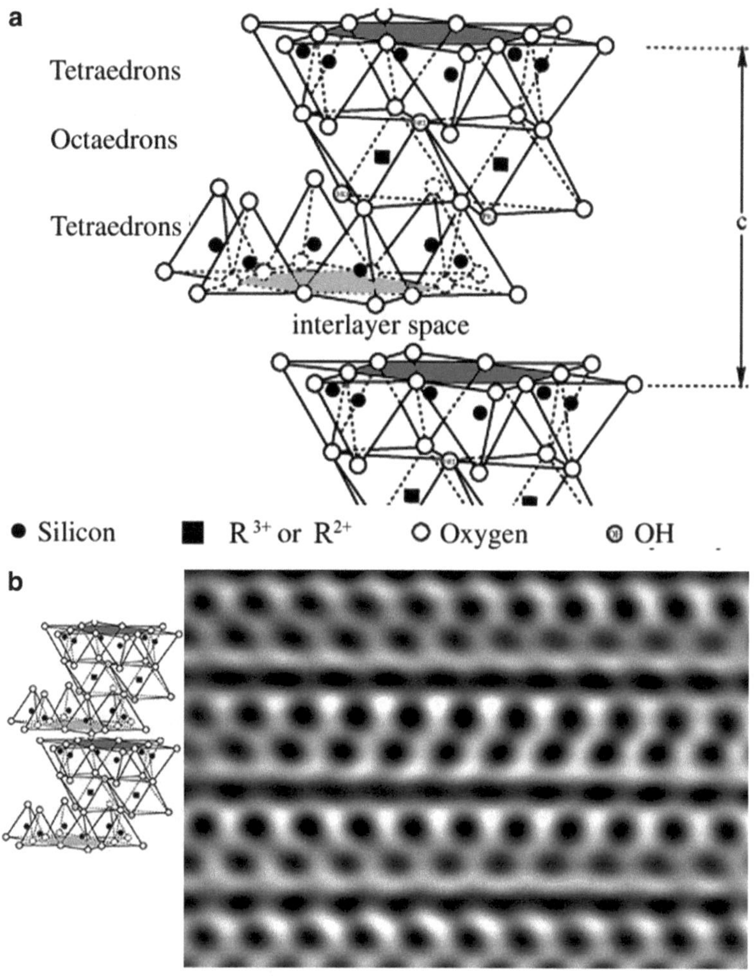

Fig. 5.6 (**a**) Structural diagram representing the structure of a TeOcTe clay mineral; (**b**) high-resolution electron microscopy image of talc whose structural diagram is represented in the image on the *left*

octahedral positions as well as compensation cations situated in an interlayer position), or talc (magnesium in octahedral position). Figure 5.6 shows this comprehensively for talc complemented with a high-resolution electronic microscopic image that allows us to visualize the tetrahedral and octahedral entities represented in the diagram. This powerful method of observation in electronic microscopy allows us to prove the reality of the arrangement of the chemical elements constituting the clay minerals' structure.

A third type of layer, a bit special, exists in the TeOcTe mode with intercalation between each layer of an additional octahedral layer, called the "hydroxide layer"[1] by mineralogists. Then we obtain the chlorite family whose structure is shown schematically in Fig. 5.7.

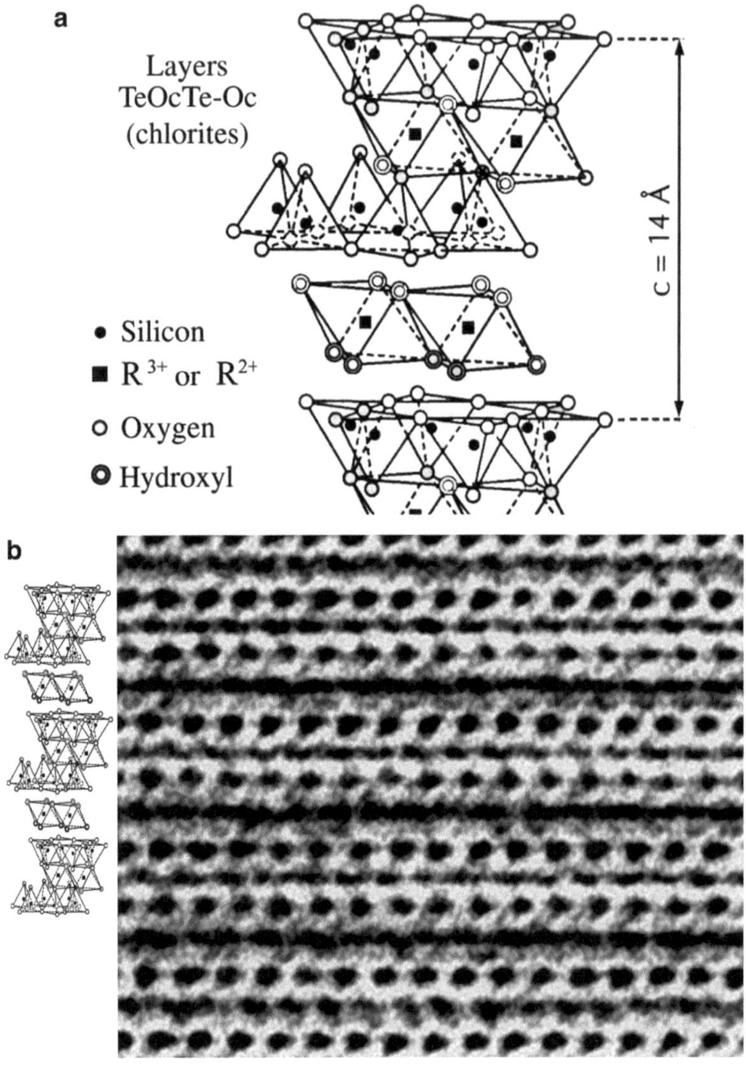

Fig. 5.7 (**a**) Structural diagram of a chlorite made of TeOcTe–Oc layers; (**b**) high-resolution electron microscopy image of chlorite whose structural diagram is represented by the image on the *left*

[1] Mineralogists qualify this layer as the "brucite layer" by analogy with brucite, magnesium hydroxide. Various atomic substitutions can exist in the layer, in particular Mg substituted by Al. The classification table and general formulas are given in the appendix.

Very high resolution electron microscopy makes it possible to visualize clay minerals' structural details; it gives an experimental visual proof of the constitution of the layers, which is very useful for nonspecialists.

These three big families present the main part of the crystalline structures indispensable to describing the entirety of clay minerals. However, it is necessary to supplement the general outline by descriptions of certain derived forms that have very specific properties but sometimes prove extremely dangerous to health (as in the case of chrysotile, a fibrous phyllosilicate and the current component of asbestos).

5.1.2 Structural Configurations of Natural Clay Minerals

The structures of the phyllosilicates that we have just defined are pedagogical and established on the basis of a geometrically perfect model, where the octahedrons and tetrahedrons constitute the framework of the structure by presenting an ideal occupation of their ionic sites. They make it possible to set up a precise classification of the various families, groups, species, and varieties of clay minerals (Rautureau et al. 2004). Tables 5.1 to 5.5 containing the classifications of clay minerals proposed by these authors, as well as Tables 5.6 and 5.7 containing the classifications of clay minerals proposed by the Joint Nomenclature Committees of AIPEA and CMS in a Guggenheim et al. (2006) report are shown at the end of this chapter.

We have already pointed out several times that natural clays contain defects in the nature and position of the ions composing them. As regards ionic substitutions, it is necessary to distinguish those localized in the octahedral sheet and those in the tetrahedral sheet.

The nature of the ions that compensate for these defects directly influences the properties of clays and particularly those of the smectites. Their charges control the local anomalies resulting from the electric charges induced by disturbances of the electric potential in the vicinity of the substitution sites. These disturbed zones are favorable for the trapping of ions and molecules from the external medium. The mechanisms are complex and more varied; it is especially necessary here to know their existence, because clay properties depend on them. For more information readers can consult Caillère et al. (1982) or the *Handbook of Clay Science* (2013).

In addition, clay particle growth is generally limited to very small dimensions: they are microcrystals, even nanocrystals. This reduced size partly results from the presence of the crystalline defects that constitute one of the essential characteristics of clays.

We take stock of these various characteristics in order to establish bonds with recognized therapeutic properties, to the extent possible.

Table 5.1 Classification based on the layer structure, the di- or tri-cation occupation of the octahedral sheet, and the electric charge per unit cell

Type of layer TeOc (2/1) Real thickness of unhydrated layer = 0.7 nm				
Dioctahedral COc = 4/6			Trioctahedral COc = 6/6	
Te = 4 Si		Te < 4 Si	Te = 4 Si	Te < 4 Si
Oc = 12/12		Oc > 12/12	Oc = 12/12	Oc > 12/12
Apparent equidistance variable	Apparent equidistance stable	Apparent equidistance stable	Apparent equidistance stable	Apparent equidistancestable
Halloysite	Kaolinite	No known species	Antigorite Chrysotile	Berthierine
TeOcTe (2/1) layer type Real thickness of unhydrated layer = 1 nm				
Trioctahedral COc = 6/6				
Te = 8 Si	Te < 8 Si			
Apparent equidistance stable	Apparent equidistance variable			Apparent equidistance stable
Oc = 12/12	Oc < 12/12	Oc = 12/12	Oc > 12/12	Oc = 12/12
Talc	Smectites Hectorite	Saponite	Vermiculite	Phlogopite
→ Increasing electric charge →				
TeOcTe (2/1) layer type Real thickness of unhydrated layer = 1 nm				
Dioctahedral COc = 4/6				
Te = 8 Si	Te < 8 Si			
Apparent equidistance stable	Apparent equidistance variable			Apparent equidistance stable
Oc = 8/12	Oc < 8/12	Oc = 8/12	Oc > 8/12	Oc = 8/12
Pyrophyllite	Smectites Montmorillonite	Beidellite	Dioctahedric vermiculite	Illite (muscovite)
→ Increasing electric charge → Te for tetrahedral sites et Oc for octahedral sites				

Coc signified "octahedral cation" and *Oc* signified "octahedral charge"

Notes: We use the nanometre, legal unity: 1 nm = 10 Å

5.1.2.1 Ionic Substitutions

There are many possibilities of substitutions; they can be arranged in three categories.

1. *Isomorphic homovalent substitutions:* The usual ions presented in the models we have just described can be replaced by ions of the same co-ordination number. The modification then relates only to the nature of the ion. Thus in the octahedral situation, Mg^{2+} can be substituted by Fe^{2+}. Because the charge is preserved, it does not have any external compensation with the layer. However, the diameters

Table 5.2 Classification of TeOcTe phyllosilicates relative to the electric charge per unit cell

Charge for cell (in electron)	Dioctahedral	Trioctahedral
0	Pyrophyllite $(Si_8)^{IV}$ $(Al_4)^{VI}$ $O_{20}(OH)_4$	Talc $(Si_8)^{IV}$ $(Mg_6)^{VI}$ $O_{20}(OH)_4$
	Smectites	
0.4–1.2	Montmorillonite $(Si_8)^{IV}$ $(Al_{4-y}Mg_y)^{VI}$ $O_{20}(OH)_4$, M^+_y Beidellite $(Si_{8-x}Al_x)^{IV}$ $(Al_4)^{VI}$ $O_{20}(OH)_4$, M^+_x	Hectorite $(Si_8)^{IV}$ $(Mg_{6-y}$ $Li_y)^{VI}$ $O_{20}(OH)_4$, M^+_y Saponite $(Si_{8-x}Al_x)^{IV}$ $(Mg_6)^{VI}$ $O_{20}(OH)_4$, M^+_x
	Vermiculites	
1.2–1.8	$(Si_{8-x}Al_x)^{IV}$ $(Al_{4-y}M^{2+}_y)^{VI}$ O_{20} $(OH)_4$ ↓ K^+_{x+y}	$(Si_{8-x}Al_x)^{IV}$ $(Mg_{6-y}R^{3+}_y)^{VI}$ $O_{20}(OH)_4$ ↓ K^+_{x+y}
	Micas	
2	Muscovite $(Si_6Al_2)^{IV}$ $(Al_4)^{VI}$ $O_{20}(OH)_4$, K_2	Phlogopite $(Si_6Al_2)^{IV}$ $(Mg_6)^{VI}$ $O_{20}(OH)_4$, K_2
4	Margarite $(Si_4Al_4)^{IV}$ $(Al_4)^{VI}$ $O_{20}(OH)_4$, Ca_2	Clintonite $(Si_4Al_4)^{IV}$ $(Mg_6)^{VI}$ $O_{20}(OH)_4$, Ca_2

Table 5.3 Classification of chlorites

TeOcTe-Oc ou (2/1/1) Real thickness # 1.4 nm $(Si_{8-x}Al_{2x})^{IV}$ $(M^{2+}_6)^{VI}$ $O_{20}(OH)_4(R^{3+}_{2x}, R^{2+}_{(6-2x)})$ $(OH)_{12}$				
Pseudochlorites	*Trues chlorites (clay facies, leptochlorites)*			
Equidistance variable Hydroxide sheet more or less complete	Equidistance stable Hydroxide sheet continue			
	Mica sheet 2/1			
	Dioctahedral		Trioctahedral	
	Hydroxide sheet		Hydroxide sheet	
Deficit de charge partially balanced by exchangeable cations List the nature of the two octahedral layers with reference to trues chlorites	Dioctahedral Dombasite	Trioctahedral Manandonite (some substitutions of Si by B) sudoïte cookeite	Dioctahedral No know example	Trioctahedral Are divided taking into account the importance of Si–Al substitution and the nature of octahedral cations constituting the layers

To define varieties it is necessary to consider the level of oxidation (substitutions Fe^{3+} to Fe^{2+}) and the succession of various layers, 1T, 2T, 3T and unordered X

Table 5.4 Classification of interstratified clay minerals

Interstratified					
	Type to 1 nm	Type to 1 or 1.4 nm			Type to 1.4 nm
	1 nm stable 1 nm variable	1 nm variable 1.4 nm variable	1 nm variable 1.4 nm stable	1.4 nm stable 1 nm variable	1.4 nm variable 1 nm stable
Irregular stacking	Hydro micas				
Regular stacking	Rectorite Allevardite			Corrensite Tosudite	Sangarite
Types 1.4–0.7 nm				Divers types	
1.4 nm stable 0.7 nm stable	1.4 nm variable 0.7 nm stable			–silica	

Table 5.5 Classification of pseudo-layer clay minerals

Pseudo-layer minerals: discontinue octahedral sheet	
Trioctahedral	Partially dioctahedral
Sepiolite $(Si_{12})^{IV}$ $[Mg_8]^{VI}$ O_{30} $(OH)_4$ $(OH\ H^+)_4$ $8(OH_2)$	Palygorskites $(Si\ Al)_8^{IV}$ $(Mg\ Al\ Fe^{3+})_5^{VI}$ O_{20} $(OH)_2$ $(OH\ H^+)_4$ $4(OH_2)$

Note: These two formulas are given for ½ crystallographic cell

Table 5.6 Classification of planar hydrous phyllosilicates (in Guggenheim et al. 2007)

Layer Type	Interlayer Material[a]	Group	Octahedral Character	Species[b]
1:1	None or H_2O only ($x \sim 0$)	Serpentine–kaolin	Trioctahedral	Lizardite, berthierine, amesite, cronstedtite
			Dioctahedral	Kaolinite, dickite, nacrite, halloysite (planar)
			Di,trioctahedral	Odinite
2:1	None ($x \sim 0$)	Talc–pyrophyllite	Trioctahedral	Talc, willemseite, kerolite, pimelite
			Dioctahedral	Pyrophyllite, ferripyrophyllite
	Hydrated exchangeable cations ($x \sim 0.2$–0.6)	Smectite	Trioctahedral	Saponite, hectorite, sauconite, stevensite
			Dioctahedral	Montmorillonite, swinefordite, beidellite, nontronite, volkonskoite

(continued)

Table 5.6 (continued)

Layer Type	Interlayer Material[a]	Group	Octahedral Character	Species[b]
	Hydrated exchangeable cations ($x \sim 0.6$–0.9)	Vermiculite	Trioctahedral	Trioctahedral vermiculite
			Dioctahedral	Dioctahedral vermiculite
	Hydrated mono-or divalent cations ($x \sim 0.6$–0.85)	Interlayer-deficient mica	Trioctahedral	Wonesite[c,d]
			Dioctahedral	None[d]
	Nonhydrated monovalent cations (≥50% monovalent, $x \sim 0.85$–1.0 for dioctahedral)	True (flexible) mica	Trioctahedral	Phlogopite, siderophyllite, aspidolite
			Dioctahedral	Muscovite, celadonite, paragonite
	Nonhydrated divalent cations, (≥50% divalent, $x \sim 1.8$–2.0)	Brittle mica	Trioctahedral	Clintonite, kinoshitalite, bityite, anandite
			Dioctahedral	Margarite, chernykhite
	Hydroxide sheet (x=variable)	Chlorite	Trioctahedral	Clinochlore, chamosite, pennantite, nimite
			Dioctahedral	Baileychlore
			Di-,trioctahedral	Donbassite
			Tri-,dioctahedral	Cookeite, sudoite
				None
2:1	Regularly interstratified (x=variable)	Variable	Trioctahedral	Corrensite, aliettite, hydrobiotite, kulkeite
			Dioctahedral	Rectorite, tosudite, brinrobertsite
1:1, 2:1			Trioctahedral	Dozyite

[a]x is net layer charge per formula unit, given as a positive number.
[b]Not an exhaustive list of species; in general, listed in order of abundance.
[c]Net layer charge may be <0.6, but this is an exception.
[d]"Series"names are given in Rieder et al. (1998) as a convenient way to describe incompletely investigated micas. For example, biotite is a trioctahedral true-mica series name for certain dark micas that may be used as a field term, and illite is a dioctahedral interlayer-deficient series name to describe certain micas after only optical microscopic data become available. Other dioctahedral interlayer-deficient micas of a series type are glauconite and brammallite.

of the initial ions and the substitution ions are different.[2] This new situation will involve distortions inside the crystalline mesh.

2. *Isomorphic heterovalent substitutions:*[3] The replacement ions may not have the same electric charge. For example, in montmorillonite octahedral Mg^{2+} is replaced by Al^{3+} which creates an electric charge defect. The most current

[2]For example, the diameter of the ion Fe^{2+} is 0.148 nm, whereas that of Mg^{2+} is 0.132 nm.

[3]They are substitutions between ions bearing different electric charges.

Table 5.7 Classification of nonplanar hydrous phyllosilicates (in Guggenheim et al. 2006)

Layer Type	Modulated Component	Linkage Configuration	Unit Layer, $c \sin\beta$ Value	Traditional Affiliation	Species
Modulated Structures					
1:1 Layer	Tetrahedral sheet	Strips	7 Å	Serpentine	Antigorite, bementite
		Islands	7 Å	Serpentine	Greenalite, caryopilite, pyrosmalite, manganpyrosmalite, ferropyrosmalite, friedelite, megillite, shallerite
		Other		None	None
2:1 Layer	Tetrahedral sheet	Strips	9.5 Å	Talc	Minnesotaite
			12.5 Å	Mica	Ganophyllite, eggletonite
		Islands	9.6–12.5 Å	Mica/complex	Zussmanite, caryopilite, manganpirosmalite, ferropyrosmalite, friedelite, megillite, schallerite, nelenite
		Other	12.3 Å	Other	Bannisterite
			14 Å	Chlorite	Gonyerite
	Octahedral sheet	Strips	12.7–13.4 Å	Pyribole	Sepiolite, loughlinite, falcondoite, palygorskite, yofortierite
Rolled and Spheroidal Structures					
1:1 layer	None	Trioctahedral	—	Serpentine	Chysotile, pecoraite
		Dioctahedral	—	Kaolin	Halloysite (nonplanar)

compensation of this charge anomaly is done outside the layer, in interlayer space, with an ion called the "compensating ion".

(a) Compensation can also occur inside the layer, in the same layer, or in the adjacent layer. In vermiculites the tetrahedral charge anomalies are compensated by octahedral opposite charges. The determination of the position of the ions that are at the origin of the charge anomalies is very complex; it requires thorough knowledge of the physicochemical properties of the mineral. This work belongs to the field of physics.

(b) According to the specific case, the compensation ions are strongly associated with the layers and they can be separated with difficulty (e.g., micas, some illites, glauconite, etc.) or they are rather slightly dependent on the layers and in that case easily mobilized and thus exchangeable or hydratable (in the case of smectites).

3. *Substitution of oxygen:* The crystallochemical models of clay minerals are established by making the assumption that all the anions constituting the polyhedrons (tetrahedrons and octahedrons) are oxygen and/or hydroxyl. This hypothesis aims for simplicity. However, the chemical analyses almost always indicate a consider-

able quantity of fluorine which can only replace for the oxygen or hydroxyls ions. Very often, by analytical facility, this chemical element is not measured out as fluorine and is thus generaly not taken into account in the calculation of the structural formulas of clays and clay minerals but as oxygen, and is thus not taken into account in the calculation of the structural formulas. In fact, this aspect is rarely approached within the framework of the crystallochemical studies of the mineral structures because it does not change the results to a significant degree. *Nevertheless, the existence of fluorine*[4] *ions cannot be disregarded, because this element has many implications with respect to the living organisms that have an essential need of it.*

5.1.2.2 Distortions of the Crystal Lattice and the Particle Shape

When a single clay mineral is observed in electron microscopy, we see that certain species have well-defined shapes, such as kaolinite, whereas others are deformed and poorly defined, such as smectites. Very quickly a rather obvious relationship is established between the existence of structural defects and crystal shapes. There is then a logical tendency to conclude that their structure and texture confer properties specific to minerals. Thus, the smectites have properties specially induced by their exchange capacity whereas kaolinite or talc are recognized for their neutral charge as well as for their crystal edge charge effect of the layers.

Imaging of clay particle deformation may be done with electron microscopy but with difficulty. The specimen needs to be frozen but that disturbs it from its original liquid state. An Environmental Scanning Electronic Microscope (ESEM) allows for minimal disturbance of a specimen, however, the availability of these devices is relatively limited.

The image of sepiolite fibers[5] from Ampandrandava (province of Tulear, Madagascar) selected for its particularly impressive length illustrates this subject (Fig. 5.8). It can easily be seen that the distortions of the crystal lattice of this fibrous mineral made up of pseudo-layers can induce tensions which can be locally important. If they are too strong, the network does not resist and the fiber breaks. The same happens with layered clays when these reach great lateral extension.

This observation must be associated with the plasticity property of concentrated clay pastes.

Thus, if the mineral can bear such distortions, it is logical to relate them to local structural properties or to variations of these properties where these curves appear; this is very delicate basic research to carry out.

The precise fitting of ions in real natural structures sometimes imposes an important differentiation compared to the ideal models we have just described. A remarkable structure having a significant impact on the properties is the rolling up of the layers in the form of tubes. The result is then a very particular phyllitic mineral: the

[4] As with some other ions such as Br^+, Cl^-, and so on.

[5] The sepiolite is made of pseudo-layers that give it a truly three-dimensional structure. A well-known variety of sepiolite is "sea foam," often carved to manufacture bowls of pipes. Still a use close to the field of health!

Fig. 5.8 Sepiolite from Ampandrandava (Madagascar) observed by electron transmission microscopy (TEM)

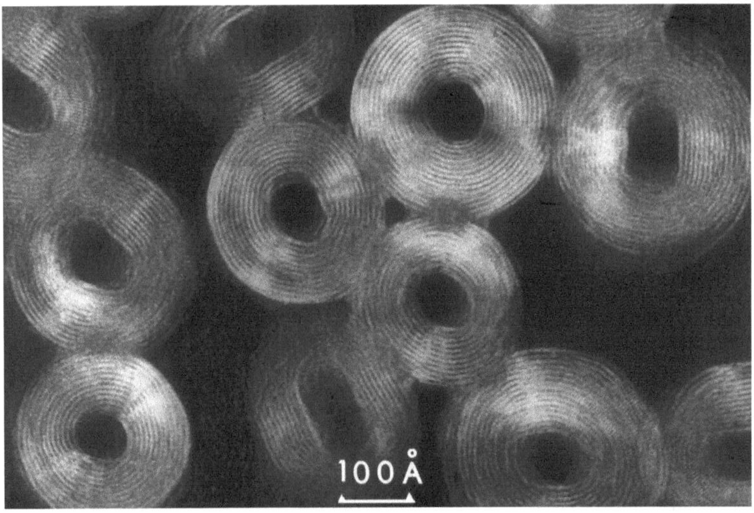

Fig. 5.9 Chrysotile tubes observed on sections perpendicular to the fiber axis. Image obtained by high-resolution electron transmission microscopy (TEM)

chrysotile that is made of hollow microtubes and presents a fibrous macroscopic aspect. A section of these tubular fibers is shown in Figs. 5.9 and 5.10 that shows their lengthening. Everyone knows the industrial product that largely uses this mineral under the French name—*amiante*—and in English—*asbestos*[8]— which is carcinogenic and contains chrysotile, a phyllosilicate (crystalline structure proper to clay minerals, strictly speaking).

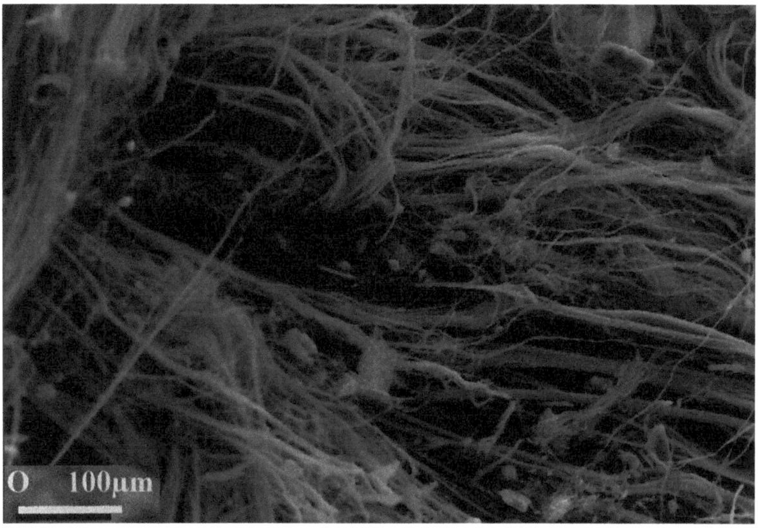

Fig. 5.10 Chrysotile fibers observed by electron scanning microscopy (SEM)

This important remark leads us to insist on the fact that clays are not automatically favorable products for health; the truth is really far from it. The "natural" origin of a product should never be regarded as a systematic factor of harmlessness.

5.1.2.3 Stability and Fragility of the Crystal Lattice and Clay Mineral Particles

Up to now we have been describing clays in a stable mineral form. We nevertheless evoked degradations likely to deteriorate them. These are frequent, because as we have specified in the foreword of this book, clay minerals are also to some extent "structures in constant evolution," if changes occur in their environmental medium.

The requests and constraints imposed by the surrounding medium can degrade clay minerals by releasing their constituents. The most current situations are the hydrolysis and acid attack that preferentially act starting from the layer edges. In other words, we say these reactions corrode the particles. To show this effect, it is necessary to produce very high magnification images using electron microscopy, which is not always easy to obtain. Figure 5.11a shows kaolinite particles whose edges still sharply delimit the unaltered layers. On the other hand, Figs. 5.11b,c show the same mineral under attacks of variable intensity; as the attack proceeds, particle edges become less and less sharp and increasingly sinuous relative to the original material. The elements that compose the released matter pass to the surrounding external medium. In a suitable context (in the case, e.g., of a water–ground–plant type system in nature or of a food–clay–intestine system with clay ingestion by a human or an animal) these elements become available for a possible transfer to the cells of the living organisms if the requirements for this reuse are met. Apart from a precise context, we cannot detail this evolution which is naturally controlled by the local physicochemical conditions.

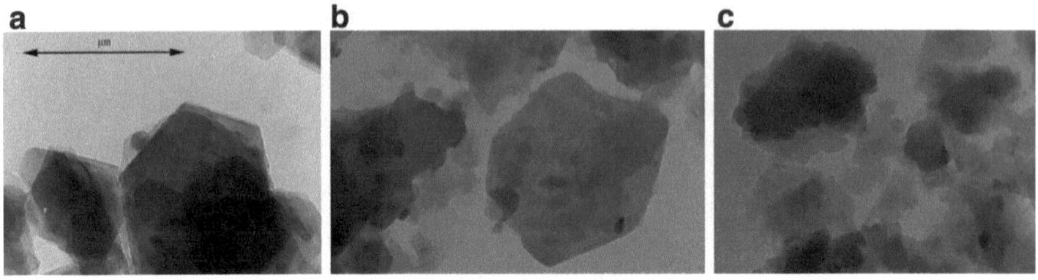

Fig. 5.11 Deterioration of kaolinite particles shown in transmission electron microscopy (TEM): (**a**) nonfaded kaolinite particles; (**b**, **c**) kaolinite particles having undergone deterioration (by hydrolysis) on the edges of the layers

Other constraints coming from extremely violent physical actions, such as intense mechanical crushing, divide the particles into smaller and often poorly crystallized fragments. The entire surface of the basal faces of the layers is practically unmodified, but on the other hand the edges of layers are redrawn, which increases deterioration possibilities.

During deterioration (acid attack, hydrolysis, etc.), the released chemical elements are almost free and independent. They are, however, quickly mobilized by the cells of the organism into new assemblages to form new assimilable compounds. We consider here only the clay particles, without taking into account the mineral impurities inevitably present.

Among the chemical elements likely to be released from clay minerals according to their composition, the following deserve to be mentioned.

1. Silicon is present in all clay minerals and is universal and dominating; it can be found in various forms of siliceous nanominerals.
2. Traditional elements of the octahedral sheet with the prevalence of aluminum, magnesium, and iron.
3. Then come the compensation elements from substitutions.
4. Finally oxygen and hydroxyl groups are in certain cases substituted by fluorine that is also mobilized in considerable quantities.

5.2 Physical and Physicochemical Properties

5.2.1 Importance of the Physical Status of the Clays Being Used

We start by treating physical properties. In effect, clays are microcrystalline solid minerals. To put them in a chemical reactivity context, it is initially necessary to make sure their interparticle spaces are accessible.[6]

[6]Let us recall, avoiding any confusion, that the clay particle in general has several layers separated by an interlayer space; these groups of named layer particles are themselves separated by a broader space, known as interparticle space, easily accessible to a fluid that remains free of all restraints compared to the mineral.

Natural clay is compact under the normal conditions of a dry layer. The large blocks are dense and hard, mechanically resistant, and not easily penetrable by fluids up to the microscopic scale where the layers have a natural tendency to set themselves in parallel one upon the other, as we have just seen. This texture explains mainly why clay layers are impermeable to water.

If one wishes to discuss reaction properties of clays, which is not really our aim here, it is necessary to reach the level of the layers. This requires crushing, then finely dividing the large blocks before using them for medical ends. The dimension of the end products can go from the small block (nearest centimeter) for rough clay delivered as "ghassoul," to the powdery form for the majority of commercial clays of the smectite type or talc, for example.

A mechanical treatment is often associated with a purification process that depends on the nature of the natural products associated with the clay mineral. Any handling by crushing can influence the continuity of the crystal lattice.[7] Thus various "qualities" of clays are marketed and it is necessary to pay attention to the description of the product, when specified. In effect, the rupture of the crystal lattice creates anomalies comparable to new board layers equipped with the properties associated with them.

Another point is the comparison of the specific reactivity to each clay family. A currently quoted case is that of the compared reactivity of a smectite and an illite, both abundantly used by suppliers. Knowledge of the structural data (atomic organization) and textural data (shape and dimension of the particles made of one or several layers) is essential to understanding that two basically different materials can have comparable reactivity: active sites are more numerous in the case of smectites but more accessible in illites (lower dimensions that give larger specific surfaces).

In clays other minerals and nonminerals are commonly associated with clay minerals. If a suitable treatment has been made in order to market a product, nearly all of them are eliminated. However, there remains a category that because of its dimensions can practically never be completely extracted: it is the majority of associated minerals present in the form of very small particles, smaller than a micrometer, even approximately a nanometer. These particles have a very important specific surface.[8] Thus, quartz and silica, inter alia, can remain in marketed products. Their properties thus will play their part with those of the clay fraction itself.

Let us take the case of silica which is absorbent because its surface ions spontaneously form connections with water. It is hydrated all the more easily as its dimensions are very small. This property is often used to eliminate water and thus to maintain anhydrous small spaces (in the case of certain conditionings of drugs in the

[7] Electronic microscopy in transmission and the diffraction of X-rays make it possible to check that after an extremely strong crushing the clay microparticles lose their crystalline character to become almost amorphous. Dispersion by very powerful ultrasound has the same effect (Gouami 1996).

[8] "Specific surface" is defined by the ratio of the surface of a particle to its mass. When a body has increasingly smaller dimensions, its specific surface becomes increasingly larger, conferring on the material a prevalence of its surface properties relative to its volume properties.

form of tablets sensitive to water); it is then useful. However, this desiccant action can appear harmful in certain uses, for example, in beauty care. The colloidal silica particles can be too drying for the skin, and excessive water fixed on silica can support the development of well-known negative effects such as silicosis. It is advisable to be careful in the choice of a commercial product and thus to refer to an experienced specialist, or require a medical leaflet including the necessary mineralogical information and the evidence of complete analyses.

5.2.2 *Importance of the Dispersing Medium*

5.2.2.1 Fluid Continuous Medium

Generally water is the dispersing liquid, to which medically active components will possibly be added. We deal with the most common simple case, that of water.

Setting clay and liquid in contact starts a process of absorption that conditions the wettability of the blocks. By capillarity, water gradually penetrates the material's porosity, that is, the open space left between the solid particles, making it possible for the mass to pass from the state of "slightly wet solid" to that of "paste." At the same time, the process of hydration of the interlayer ions can gradually develop. The hydration of the medium stops spontaneously when the paste becomes impermeable. A mechanical action is then necessary to allow water penetration and to pass in a fluid, called "mud" and also called "*barbotine*" in French by specialists in ceramics and "gel" by physicists. We are in the presence of two types of water that have very distinct particle binding energies.

The mixtures obtained are malleable or fluid; their properties then obey the principles of thixotropy and rheology. They are the properties of a macroscopic set of particles. It is very difficult, at least in the state of current theories, to define a single model that makes it possible to explain the passage of specific microparticle properties with those of particles in partnership with water. Consequently, we consider that these two scales are distinct.

We do not develop overly precise concepts here that can be consulted in other more specific works, if necessary. However, a very simple analysis allows us to estimate and understand the importance of the force of the connections between water and clay in the various situations we have just examined. This method is thermogravimetric analysis: the sample is weighed while its temperature is gradually increased. We note weight losses; here water is lost in successive steps. Distinct energies are needed to cause the withdrawal of the various categories of water molecules: the binding strengths with the structural framework are higher when the temperature of these withdrawals rises.

Figure 5.12 shows the curves of thermogravimetric analyses obtained for various clays. These curves indicate the weight loss (here due to the loss of water) according to the temperature. Recording the entire range of temperature is necessary to deter-

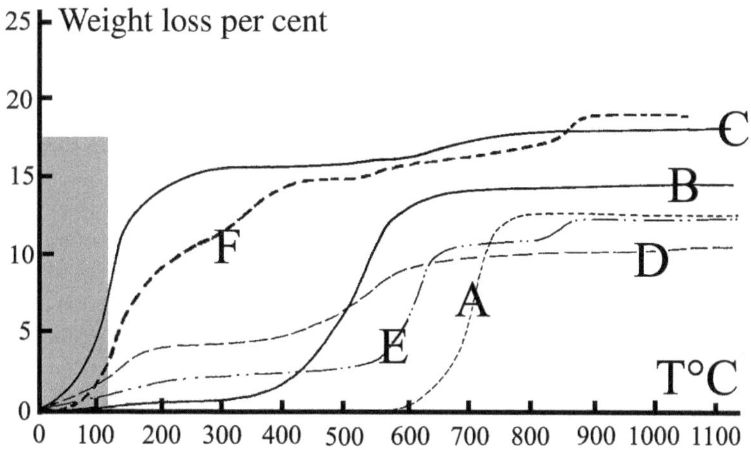

Fig. 5.12 Thermogravimetric analysis curves of various phyllosilicates: *A*—antigorite; *B*—kaolinite; *C*—montmorillonite; *D*—illite; *E*—chlorite; *F*—sepiolite. Note: Only the zone located between 0 and 100°C (zone grayed on the diagram) is interesting for health purposes

mine the type of clay studied. Therapeutic clays are employed at room temperature (or close to it) in order to be supported by the organism. In this case, the only mechanisms in question are in connection with interstitial free water. However, it is absolutely necessary to know the other types of water. In effect, if the water is mobile according to thermal energy, it is the same for chemical or physics interactions that bring the water molecules near certain reagents. The energy transfer can cause a rupture of chemical bonds and a new assignment of the roles of the different components. In this reaction pathway, the interlayer ions have a key role because they can become active again; the ionic exchange mechanism is an application of it: it intervenes, for example, in the use of cataplasms.

Many methods give access to the binding energies that constitute a solid basis for knowing the mineral and the state of the real structure in the associated aqueous medium and thus to consider its reaction behavior.

At low temperatures (definitely lower than 100°C) used for body contact, we observe that the phenomena which develop are reversible and do not destroy the mineral's crystalline structure. The thermal analyzed curves depend on the state of the clay (blocks, dry powder, mud, paste, or suspension) and reflect connections between the clay mineral and the chemical elements associated with it. They are very useful in understanding the reaction mechanisms.

For low temperature withdrawals, the water associated with clay is not very dependent. Thus we find interstitial water or free water located between small crystallites, the water bound by weak hydrogen bonds, and finally the zeolitic water that occupies some small empty spaces of the structure (vacuoles due to deformations of the layers; central channels of rolled up fibers of halloysite and chrysotile; phyllosilicate channels with pseudo-layers of sepiolite or palygorskite).

5.2.2.2 Gas Continuous Medium

Air is the most common continuous gas medium, often wet, containing water in the form of vapor. In this case, clay is in an almost dry situation and contains only ambient residual moisture. Water is then partially related to the structure (edges of the layers) or trapped in porosity.

This situation is not exceptional; talc powder that we use daily is in this state. The properties are then those of the particles themselves. The layers of talc, magnesium phyllosilicate, are electrically neutral; they have no surface forces of attraction between the particles. With the slightest effort they slide softly over each other. This property of sliding is used, for example, in physiotherapy to avoid irritation due to the friction of the hands on the patient's skin.

We have just introduced a property using the interactions between particles. We can also observe the particles themselves. Their low dimensions can give some of them abrasive properties when they are plunged into air or water (field of tribology). That can seem contradictory to the better-known properties of plasticity of wet grounds. Actually, they are two properties that utilize scales of very different sizes. The kaolinite microparticles, for example, are very rigid on a micrometer scale; they typically present crystallized forms, that is, with faces and edges well defined (Fig. 5.11). This property is largely used in cosmetic and aesthetic care for skin gumming and exfoliation, for example.

5.3 Various Types of Water

The essential thermal properties of water are well known. Pure water freezes at 0°C and boils at 100°C; these reference temperatures change if water contains dissolved elements. Between the two, it evaporates slowly, starting from its free surface if it exists and if the external medium allows it. The connections that it can contract modify these figures.

When interested in the reactivity of water in a water–clay system, it is thus necessary to differentiate several types of water.

5.3.1 Interstitial Water

Interstitial water or free water is free of any connection (with a clay particle, e.g.). The connections that it can develop modify the figures given above.

The water located between the clay layers is almost free but all the same, evacuates itself completely only above 100°C which is usual for free water. It is a slightly dependent zeolitic-type water, that is, by hydrogen bonds, for example, or being in cages such as the sepiolite channels. In all cases, molecular water is always completely released as soon as the temperature reaches approximately 105°C.

Any withdrawal at higher temperatures implies that water has stronger connections with minerals or that this water enters the structure itself (water of crystallization and hydroxyl structural groups). These types of dependent water are implied in the behavior of pastes and oxilic water. For example, in the case of calcic clay (containing calcium as interlayer compensation cations), water withdrawal is between 120 and 125°C.

5.3.2 Water Integrated into the Structural Layers

Let us return to clay minerals themselves. We previously specified that they are hydrated phyllosilicates. Their crystalline structures contain internal hydroxyl ions within the layers; they are known as "structural." In Figs. 5.5 and 5.6 these hydroxyl groups are denoted by the symbol OH^- located in a circle. The structural balance of the unit cell imposes the presence of these hydroxyl ions which are thus "structural ions." They are very strongly related to the octahedral cations of the layer and can be eliminated only by one strong rise in temperature, which is not the case for medical uses.

However, important variations of pH in the stomach medium, for example, can deteriorate the structure almost completely and thus move these OH^-.

5.3.3 Water Integrated into the Structural Layers

We know that clay minerals are made up of small particles of less than two micrometers constituted of layers forming packages similar to a book. The particle has a contact surface with the external medium in which it is contained. The external surface of the particle (made of a higher layer surface and another of the lower layer as well as the side surfaces of the layers) is directly accessible to the liquid. The charges of the ions that are on the edge of these layers are not compensated and cannot remain in this situation. One of the most current cases of neutralization of these electrical charges is the fixation of H^+ ions or OH^- (of which we give one possible configuration in Fig. 5.13). This makes it possible to have a great activity of water fixed by hydrogen bonds on the layer edges and infinite possibilities of later bindings.

5.3.4 Hydration Water of the Interlayer Cations

Heterovalent ionic substitutions internal to the layers are at the origin of electric anomalies of charges (with defects and excesses) compensated in interlayer space by ions of opposite charge that can join water molecules known as "hydration water" solvating the interlayer cations.

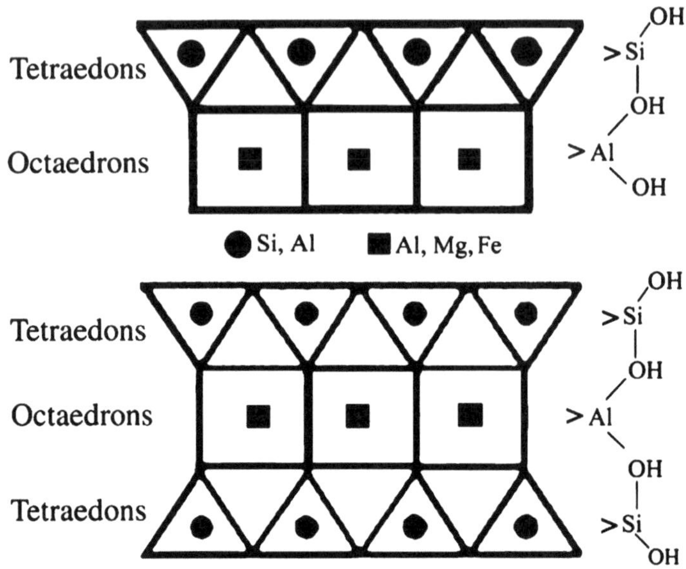

Fig. 5.13 Schematic representation (on the *left*) of the edges of TeOc and TeOcTe layers important to supplement and assure neutrality

The spontaneous behavior of these products depends on the importance and accessibility of the contact surface between the particles and continuous medium, which is generally air for dry products.

In the presence of water, the interlayer cations can be hydrated gradually by fixing a growing number of layers of water. The clay properties evolve with the number of interlayer water molecules (see Fig. 5.14). We can easily envision that the attraction forces between close layers decrease while the space between them is growing: water forms a screen which is, starting from a certain thickness, sufficient to release the adjacent layers of the mutual constraints that maintain them closer. Progression of this mechanism is measured by absorption isotherms and we note the step by step swelling of the material related to the variation of the distance between the layers.

5.4 Various Hydration States

Thermal spring water has a more complex action. This water is charged with anions and cations inherited from the various rocks through which it circulated. It contributes not only to the hydration of clay layers, but also allows other chemical exchanges.

In the presence of a liquid, clay reacts according to its ionic exchange capacity. In a very general way, this exchange capacity is important for the cations and

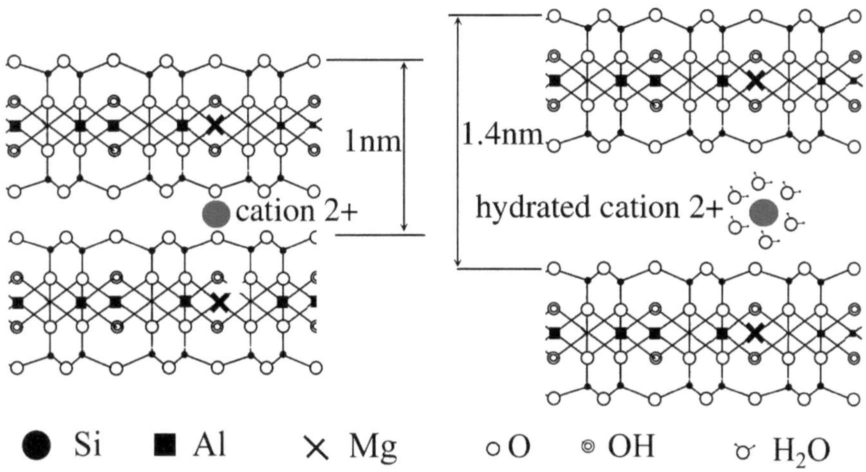

Fig. 5.14 Schematic diagrams trying to explain the swelling caused by hydration of an ion compensating the interlayer located opposite the charge deficit of the adjacent octahedral layers (in this case, aluminum substituted by magnesium)

much weaker for the anions, meaning that if clay is placed in a thermal spring, the concentration of the anions in the solution does not vary significantly. At the same time, some cations of the solution can pass in the clay's compensation position or be exchanged with those already in compensation position. These mechanisms are well known but very complex because the local situations are very diverse. Even if they are difficult to study, use in a thermal establishment must take them into consideration.

Table 5.8 gives an idea of the average cation and anion exchange capacity[9] of various clay minerals. It is noted that the anions are poorly exchanged in the case of current commercial clays. It should be remembered that the development of therapeutic activities envisaging the exchange of anions preferentially leads to the use of synthesis minerals easier to control in this field. However, the exchange capacity imposes properties that cannot be developed by ignoring the textural properties and the access complexity of the material's active ionic sites. It depends on many parameters such as compactness, tortuosity, temperature, dilution, and so on, which are to be taken into account in the product applied to the patient at the time of therapeutic use. These parameters describe and actually manage the conditions of accessibility to the reactive sites. All these characteristics influence the realization time of the exchanges and perhaps make it possible to understand their effective blocking by infinitely slow kinetics. Under this extreme condition of inefficiency, the product must be mixed again, modified, or renewed.

[9] Classically, the exchange capacity is expressed in milli equivalent grams for 100 g of clay. In the case of cations it is named CEC (cation exchange capacity); currently, it is expressed in Cmol/kg (centimole$^{(+)}$/kg).

These measurements are carried out on "dried" clay previously heated up to 100°C (up to 125°C if it is clay-Ca). Thus, all the water that does not belong to the layers is eliminated.

Table 5.8 Some values of cation and anion exchange capacities (in Caillère et al. 1982; Rautureau et al. 2004)

	Mean Exchange Capacity (in meq gram for 100 g of calcined clay)	
Mineral	Cations	Anions
Kaolinite	3–5	7–20
Halloysite	5–10	
Montmorillonite	80–150	23–30
Hectorite	44	
Palygorskite	20	
Illite	10–40	
Vermiculite	100–150	4
Chlorite	10–40	
Sepiolite	20–30	

Natural clays are almost always polycationic; it is the result of their genesis. Under such conditions it is not easy to find identical clay layers, while at the same time the security standards regarding health, however, require reaching that point. By chance, very large rather homogeneous layers exist but they are not inexhaustible. Consequently, a time will come when for significant amounts, a direct intervention on the structure of clays will be necessary. For the moment such an intervention is not to be considered. To make a monocationic clay, it is necessary to carry out successive exchanges in a solution containing the ion one wishes to use, for example, sodium, calcium, or potassium.

We insisted a great deal on the privileged relation between clay and water without entering a precise description. We must pay this relation particular attention here.

The behavior of clay changes considerably according to the water content. In fact, we should no longer speak about clay in a strict sense but rather of the water–clay mixture and consider its state of hydration. For low levels of hydration, thermal analysis can contribute to specifying the state of the clay and thus to improving comprehension of those phenomena. For the definitely hydrated states (mud, gel, and suspension), the facts are more complex and require technical concepts familiar to specialized laboratories.

The skin or the mucous membranes of a living organism are likely to be in contact with clay. The states that clay can present (dry powder, paste, or mud) generate answers quite different from the zone of contact. Many ionic or molecular settings come into play implying various energy values. There are no quantitative studies in vivo in this field yet. However, general studies can be used for the analysis of these mechanisms, particularly of adsorption and desorption, transfer, and mechanical effects.

5.4.1 *Dry Clay*

This state is delicate to define because even the concept of clay implies the presence of water.

The phrases "dry clay" and "anhydrous clay" should not be confused. Clay that has become completely anhydrous has lost all its structural water molecules and structural hydroxyl functions: it is no longer a clay but an anhydrous product said to be of "high temperature." This point is beyond the frame of the present work. Therefore, in any case, only dry clay still contains bound, or not completely free, water that belongs to the crystalline structure of minerals.

When clay is naturally dry, it is generally in the form of hard solid blocks or, after conditioning, more or less fine powder. The occupied volume is very sensitive to the quantity of water (dependent or free). A proof can be found in a very well-known phenomenon: cracks in clay soils that appear in the dryness period. The cracks in the ground are due to the lack of water. However, under normal atmospheric conditions, the temperature never reaches that of the boiling point of water which could justify a fast withdrawal. In addition, when clay is hydrated, its volume increases: it is said that it is "swelling." These two completely opposite aspects of the behavior of clay are moreover highly important in understanding its practical applications.

When all the water of the clay has been eliminated, the rehydration is not immediate; it can be slow, even very slow. The phenomenon is reversible but with different kinetics (start and return speeds of water). A dry clay is a complex building of microcrystalline particles allowing a porosity (macroporosity compared to the size of the particles) to appear, responsible for a strong capillarity. Water undergoes the effect of capillarity and penetrates the mass by macroporosity. We might think that hydration is finished but it is not the case because a nanoporosity remains constituted by the interlayer space of clay (see Fig. 5.3).

This great complexity of the hydration mechanisms also applies to the hydration of the compensation ions. The speed of progression of water is not similar in all interstices. It is one of the reasons for the well-known practical stage of "maturation of the peloids," which makes it possible to pass from the dry clay state to that of uniformly hydrated pasty clay (Veniale et al. 2004).

Clays are practically never used in a dry state. We could think the same of applications such as exfoliation of the skin, but the desiccating effect is in general too strong: the affinity of clay for water can lead to dehydration of the skin.

There remains a universal use which is that of talc. The chemical constitution of talc practically does not comprise defects: it does not present electric compensable charges. In fact, the layers are not attracted by electrostatic forces and can slip easily over each other with the least request. Talc is soft and facilitates gliding by attenuating friction, justifying its current use for body care.

5.4.2 *Clay in Paste*

For health care, paste often prepared with water is used primarily for its plasticity which enables it to adapt to the shape of the treated body. Clay particles contained in the paste are rather strongly related to one another but without the whole structure being "completely fixed" akin to a solid block. During the application of a force the deformation of the paste causes friction between particles. (It is characterized by the index of plasticity.)

The introduction of polar molecules into a clay paste influences its plasticity considerably. The contribution of water is the first example, most important in the field we deal with but many other molecules are used. Their influence on the organism must be carefully controlled.

All clay mineral properties are found in the paste and reflect in particular the presence of the interfoliaceous compensation ions. The stability of their state and the permanence of their position in the clay crystalline structure obey the rules governing traditional chemical exchanges and their balances. They confer on clay its capacity of anion and/or cation exchange: these ions can be exchanged by others respecting total electric balance. This situation is at the origin of the use of the pastes for many treatments.

To give an idea of the behavior of clay in a situation where an exchange of ions, sodium, and calcium is possible:[10]

$$\text{Smectite}\quad \text{Ca} + 2\text{Na} \Leftarrow\Rightarrow \text{Smectite}\quad 2\text{Na} + \text{Ca}$$

This reaction allows reversible ionic transfers between the body and the paste. All kinds of ions can be considered, which makes it possible to approach a great number of pathological situations by so-called "soft" therapies. The control of these mechanisms is not, however, easy, and empiricism or experience still plays an important part. Nevertheless the diffusion of scientific knowledge gradually allows better comprehension of these mechanisms, and allows increasingly more rational explanations.

When a paste is desiccated it retracts until the well-known cracks appear as in the case of dry clay soils. When a peloid is applied directly to the skin, side forces can result from this drying; a close eye has to be kept on what is happening because too long an exposure would cause damage due to dehydration of the paste involving that unsought-for the skin.

5.4.3 *Clay in Suspension*

Some applications in the field of health require clays with a great quantity of water. As mentioned above, we pass then from the concept of paste to that of suspension. However, the situation is not simple because the clay particles have surfaces that support the compensation ions. Some have an aptitude for being hydrated in more or less complex ways. These structures are built starting from the surface of the clay and induce anomalies of local electric potential that imply the existence of attraction or repulsion forces according to the nature of the charges. The passage of the paste

[10] The notations Clay-*Ca* or Clay-*Na* mean that the internal charges of the layer are compensated by calcium or sodium ions in interlayer positions; they are called, respectively, *calcic* or *sodium clay*.

or viscous mud stage to that of fluid suspension can be extremely fast, a calamitous change of physical status.

A very interesting state of the strongly hydrated water–clay systems is that of gel. Many types of clay, and particularly smectites, have the aptitude to form a gel that can possibly be used to transport ions or molecules useful for health, such as for traditional stomachic bandages containing clay. Those have primarily the role of "load" or "plug". This gel can in certain cases be formed even with very small quantities of clay.

Many suggestions were made by distributors to carry out clay suspensions in water that can then be elutriated. The liquid floating on the surface is often called "clay milk"; it is then recommended as a drink with various virtues. Analysis of this liquid is very complex. It is formed by two types of components: clay particles, isolated layers or small agglomerates of layers, which are found in suspension and cannot leave this state because they are in electric balance with the liquid, and a multitude of ions coming from clay and found in the suspension. These are components that can be active but, to explain the nature of this activity, it is necessary to know their chemical composition precisely.

In practice, in laboratories a method of decantation is employed to separate the various phases (mineralogical and dimensional) of which natural clay is always constituted. For that a suspension (not gelled) of clay in a liquid, generally of water, is carried out. This suspension is placed in a separating funnel that is kept away from any mechanical or thermal disturbance. Particles present in the liquid fall towards the bottom of the funnel more or less quickly according to their mass. We thus find a stacking of layers containing the various "mineral classes" present. They can be recovered for studies specific to each of them. On the other hand, there exists a fraction that never elutriates and remains in suspension; it is that which interests us under the name of "clay milk".

5.5 The Color of Clays

The color of clays is often evoked during their marketing and constitutes a parameter of choice for their use. What does the color of these minerals represent and what is its relationship to their properties or their purity? Dependent on the first visual perception that one has of clay, the color is often used in connection with its properties. In a certain sense this proposal is acceptable, but it is not always supported by scientific reality.

5.5.1 Importance of Color

The color of clays is a very important parameter in the applications we deal with here. There is a very great visual incidence of color on the relation between a person and a product used for personal body care. Commercial qualifiers cannot be more explicit: thus one usually sells green or white or brown clay, and so on; the mineralogical and crystallographic characters follow when they are known. In this case the color is a criterion of a primarily visual reference mark for the purchaser. Often this color criterion is almost a parameter of quality but many users cannot explain the reason for their choice.

What does the color of these minerals represent and what is its relationship to their properties? This delicate subject deserves an all the more detailed approach as it relates to a very personal phenomenon of perception. A great many works deal with this question in the most varied fields; for more general information readers may consult Zuppiroli and Bussac (2002).

Many types of clay have a white color. They are then in general required because of the absence of ion dyes that could disturb the processes for which they will be used.

The white color is often synonymous with purity, which justifies its use in the field of health and beauty care. White clays are often minerals close to the ideal structural models undisturbed by defects at the time of their genesis. This characteristic requires a great uniformity of the initial formation conditions. They can be found in large deposits (kaolinite, talc, smectites, e.g.) or in microscopic transformation sites in bedrock (kaolinite of Saint-Austell in Cornwall).

White is reassuring because it testifies to cleanliness; green is associated with calm, nature, balance, and rest. It stimulates spiritual activity; it is the sign of hope, even of immortality. Progressively all these characteristics have been embedded in the psychological depths of people. There is no doubt that the relationship between the color of clay and its use has a psychic dimension. If it is a positive contribution, why should it be rejected?

5.5.2 *Origin of the Color*

Note: Here we approach only the case of clay that is considered sufficiently pure that its color depends only on itself. In the rather frequent case of mixture with other organic minerals or molecules, the color can result from their presence (as is the case, e.g., when an alumina or iron oxide is present).

The color of clays results from their crystalline structure in a given state and context. It is thus their chemical composition, indissociable of the crystalline structure in question. The state of oxidation of the structural cations is a crucial factor of the "original" color. To understand these colors, it is necessary to associate them with the state of ionic charge and their position (tetrahedral, octahedral, or interlayer compensation ions) in the structure.

A very low ionic level or degree of substitutions is sufficient to disturb the color of white clay. For example, negligible iron traces color kaolinite (but also kaolin)[11]

[11] Two very distinct situations lead to coloring. In the case of "kaolin rock," very fine particles of oxides or metal hydroxides intervene in mixture with the other components. For kaolinite, internal substitutions for the crystalline structure can exist simultaneously with the small particles external to the structure.

towards an ochre color and consequently considerably minimize its industrial use for paper coating. In the same way, white bentonite is sought for ceramics during liquid paste manufacturing because it does not disturb the color with cooking. These small particles in the field of health, sometimes of nanometric sizes, mixed with clay, can have an undesired incidence, thus it is appropriate to be careful.

In the field of clay synthesis, control of the chemical composition allows mastery of the color. Thus "talcs of various colors" were obtained in 2007 by F. Martin of the Laboratory of the Mechanisms and Transfers in Geology, CNRS/IRD Toulouse-III University, by introducing specific ions in octahedral positions, cobalt (pink), nickel (green), or copper (blue). Their introduction into paramedical applications is thereby possible.

For a more complete scientific approach to this question, readers may refer to the work coordinated by Decarreau (1990). It must be said that a clay extracted from its natural environment can evolve slowly by change of the oxidation state of certain ions under the atmospheric conditions of temperature, pressure, and hygroscopy. Thus samplings of clays can lead to erroneous identification conclusions (e.g., halloysite is dehydrated quasi-spontaneously; glauconite and nontronite turn progressively red by simple exposure to the sun because of the change of iron's state of oxidation).

A significant amount of water can completely modify the color aspect. The presence of water, even in a small quantity, modifies the color aspect of clay and it becomes brighter and more saturated.

The origin of color of clays is not unique. It is often in the type of structural defects, but it can also come from matter associated with clay, and reflect the presence of chromophoric functions and colored centers.[12] The presence of associated compounds can influence the visual aspect considerably, and in certain cases, it is advisable to remain very vigilant, because the color can come from impurities whose properties cannot be easily examined.

Removing structural ions from silicate layers within the mineral is very difficult in practice, even impossible. The natural or forced modification of their state of oxidation influences the electric potential and thus causes a change of their characteristics with respect to the visible spectrum of their color. Heat often modifies the color; however, it should not be forgotten that, in the case of high temperatures used by the ceramics industry, one gets very different types of structures compared to those of the original clays; it is not the subject of the present work.

Clays contain a second type of ions, known as compensation ions, described previously. They are located in interlayer positions or on the edges of the layers. Their connections with the layers are much weaker; they are thus easily exchangeable. Their action on the color is important and more or less controllable. They allow many applications in the research of visual effects.

The combination of effects of the structural and compensation ions on the color can lead to difficulties of interpretation at the time of mineralogical determinations (Table 5.9).

[12] Certain functional groups of atoms are colored; the color can also come from local electronic defects.

Table 5.9 Color of clay minerals depending on the nature of the structural ions

Colorations of Clay in Natural State	Structural Ions
White	Al^{3+}, Mg^{2+}
Blue	$Fe^{2+}, Cu^{2+}, (Ni^{2+})^{IV}$
Yellow	Fe^{3+}
Rose	Li^{+}, Mn^{4+}, Co^{2+}
Red	$Mn^{3+}, Fe^{3+}, Co^{3+}, Ti^{4+}$
Green	$Fe^{2+}, Fe^{3+}, Cr^{3+}, Ni^{2+}$
Violet	Mn^{2+}, Cr^{2+}

Note: This table gives only essential indications; it is not exhaustive.

Geologists and mineralogists always locate the colors in the field, because the lines of isocoloration in general testify to similar phenomena of evolution during or after formation. These observations can be made directly in the field or in samples. The homogeneous zones and the intermediate transition zones can thus be separated and differentiated to get a first approximation of the local chemical composition in minor elements and differences of the oxidation step between zones. From one place to another, it is then possible to measure the variation of composition and, possibly, to establish its history. This stage is essential in the selection of useful clays for health.

With respect to patients, the use of the color of clay by those in charge of health concerns an extremely complex field that requires great caution.

5.5.3 *Evolution of the Color*

Two physical parameters have important effects on clays: the temperature and the state of hydration. We consider only the conditions of temperature adapted to the living world.

The color of clay changes when it is heated gradually. At a low temperature, we have seen that water begins to separate from the mineral. When the temperature rises, the oxidation changes of some ions (often iron) cause irreversible color changes. The color is a witness to the mineral history. In fact, the rise in temperature accelerates only the phenomenon that can occur at room temperature in adapted circumstances (dehydration or after dispersion, e.g.), but over definitely longer periods of time. For example, the passage of Fe^{2+} to Fe^{3+} in the octahedral layer of a smectite preserves its clay character.

Nevertheless the strongly heated products—even if the structural framework of the initial clay remains—can no longer be considered as clays because they have become dehydroxylated.

Color also evolves with the state of hydration. For low levels of hydration the color becomes brighter. For strong rates, for example, for a suspension containing small quantities of clay (a few percent) the clean color of clay is replaced by that caused by the phenomenon of diffusion of light on the microparticles which gives an opalescent aspect.

A good observation of the color of clay should be conducted under conditions of rather strong moisture. The colors are then brighter. That does not mean in any way that they are more representative, but that the contrasting effects are enhanced.

5.6 Classification of the Phyllosilicates

We present some simple tables that the authors hope may be helpful to the greatest number of readers. Tables 5.1 to 5.4 were extracted from Caillère et al. 1982 and Rautureau et al. 2004 and correspond to an historical teaching approach of the classification of phyllosilicates based on three parameters: layer structure, electric charge per unit cell, and di- or tri-cation occupation of the octahedral sheet.

For readers who would like more up-to-date and official information, it would be useful to consult the report showing the official nomenclature and classification of clay minerals (hydrous phyllosilicates) produced by the Joint Nomenclature Committees of AIPEA and of the CMS.

"Summary of Recommendations of Nomenclature Committees Relevant to Clay Mineralogy": Report of the International Association Internationale pour l'Étude des Argiles (AIPEA), Nomenclature Committee for 2006, by Guggenheim et al. (2006), published in *Clay Miner* 41:863–877 and in *Clays Clay Miner* 54:761–772. This report has undergone minor changes, disclosed in a later report published in *Clay Miner* 42:575–577 (2007). For comparison with Tables 5.8 and 5.9 that have been extracted from this report exhibited here, although showing a larger number of clay mineral species they are based on the same three parameters referred to earlier: layer structure, electric charge per unit cell, and di- or tri-cation occupation of the octahedral sheet.

References

Caillère S, Hénin S, Rautureau M (1982) Minéralogie des argiles, 2ème édition, 2 tomes. Masson, Paris

Decarreau A (1990) Matériaux argileux, structure, propriétés et applications, éd. Société Française de Minéralogie et Cristallographie et du Groupe Français des Argiles, Paris

Gouami A (1996) Des sepiolites sinifiées aux gels de silice fibreuse propriétés physicochimiques texturales et rhéologiques. Thèse, Université d'Orléans

Guggenheim S, Adams JM, Bain DC, Bergaya F, Brigatti MF, Drits VA, Formoso MLL, Galán E, Kogure T, Stanjek H (2006) Summary of recommendations of nomenclature committees relevant to clay mineralogy: report of the association internationale pour l' Étude des argiles (AIPEA) nomenclature committee for 2006.

Guggenheim S, Adams JM, Bain DC, Bergaya F, Brigatti MF, Drits VA, Formoso MLL, Galán E, Kogure T, Stanjek H (2007) Clay Miner 42:575–577

Rautureau M, Caillère S, Hénin S (2004) Les argiles, 2nd edn. Société de l'industrie minérale, Paris, ISBN 2-904845-30-5

Rieder M, Cavazzini G, D'Yakonov YS, Frank-Kamenetskii VA, Gottardi G, Guggenheim S, Koval PV, Müller G, Neiva AMR, Radoslovich EW, Robert J-L, Sassi FP, Takeda H, Weiss Z, Wones DR (1998) Nomenclature of the micas. Canadian Mineralogist 36: 905–912

Veniale F, Barberis E, Carcangiu G, Morandi N, Setti M, Tamanini M, Tessier D (2004) Formulation of muds for pelotherapy: effect of "maturation" by different mineral water. Appl Clay Sci 25:135–138

Zuppiroli L, Bussac MN (2002) Traité des couleurs. Presses polytechniques & universitaires romandes

Chapter 6
General Information on Clay Applications for Health and Well-Being

Abstract This chapter reports on the ordinary clay applications in the field of health: first, the function of the nature of the pathologies to be treated, second, the function of the clay potentialities to treat certain pathologies, and third, the function of the principal reactivity domains pertaining to clay and living organisms.

Adsorption and absorption are the clay properties that most justify both the ingestion of clay (in powders or water dispersion) to treat internal health disorders of the gastrointestinal tract promoting the neutralization of stomach acidity or the fixation and posterior elimination of toxins or other toxic compounds or even the supplementation of deficient bioessential chemical elements (e.g., Fe, Zn, Ca, Mg) and the cutaneous applications of clay for skin care purposes. Heat capacity is another property related to clay pastes when, for example, mud packs are used topically to treat some skeletal–muscular rheumatic pathologies and skin inflammatory disorders (e.g., psoriasis). Clay reactivity and the specificities of the distinctive modes of application and action that can occur in clay–human body interaction, emphasizing various effects, such as the barrier effect, the fixation and transfer of essential or toxic chemical elements, as well as heat transfer are dealt with as well.

6.1 Principal Domains of Clay Application

There are several ways to approach this subject:

1. We can use a classification according to pathologies or of types of care; it is the method generally found in papers concerned with alternative medicines.
2. We can define the potentialities of a clay vis-à-vis pathology.
3. We can also index the large reactivity fields specific to clays and living organisms to find the reciprocal interactions based on the concept of balance.

There is much written that would lead us to believe that "clay" is a universal answer to all sorts of health problems, which is particularly naive and sometimes erroneous. In fact, as has already been specified, the term *clay* indicates a group of minerals comprising a large number of species and varieties (the so-called *clay minerals* being fundamental constituents) that all have specific properties. In some cases, it is obvious that clay does not cure the body, but helps it to overcome a difficulty, and it is the organism which cures itself. Moreover, clays do change with

M. Rautureau et al., *Clays and Health*, DOI 10.1007/978-3-319-42884-0_6

time, or according to the way they are used. While trying to adapt to the conditions of the medium that surrounds them, they can acquire distinctly different properties from those of the initial clay mineral. In general, vague notions of the real nature and origins of the products being used[1] are found in the literature and medical prescriptions or the packaging of commercial products. This uncertainty should not be allowed to persist. This absence of information makes it impossible for medical people to know the precise reasons for therapeutic successes or failures and should be denounced. Maintaining vague notions verging on obscurantism is open to criticism; it is even unacceptable.

It is now necessary to insist on the importance of a "scientific" approach that makes it possible to define the basic foundations of knowledge of clays. To this end, remember that the advance of knowledge only very lately has allowed comprehension of the crystallographic fitting of clays, and consequently that of their mineralogy.

The authors of this book are much indebted to Caillère in regard to mineralogy starting in 1930, followed by Millot in regard to geology starting in 1948, for their fundamental works on the subject of clay. These two authors made a decisive step in the knowledge of the phyllosilicates, more commonly called "clay minerals." Their works remain recognized references, inescapable reference works for researchers and models for students.

Before going into scientific detail, let us recall the principal known uses. We treat health here and human well-being but it is obvious that the animal world is also concerned with this practice; animals have recourse to it instinctively and veterinary surgeons use it successfully. We could extend certain conclusions to plants; however, this field, which concerns pedology[2] and botany (Hénin et al. 1969; Duchaufour 1997), is not dealt with here.

Water is associated with clays at all stages of their existence and uses. Because of this very great affinity with water, clays have always been close to hydrotherapy activities. Before such use, thermal waters circulated for a long time through important masses of porous and fissured rocks. Most "mineral water" has been in contact with clays. This water, carrying ions coming from the degradation of rocks, has been used for centuries in thermal treatments. Practically, they are often added to clays for the preparation of pastes used in the preparation of peloids. Hydrotherapy experts also use clays in the form of muds to cure rheumatism, arthritis, various traumas, and skin diseases.

Kaolinite, smectite, illite, talc, sepiolite, and palygorskite are employed for their particular properties of adsorption and absorption. They make it possible to make stomachic bandage, cutaneous treatments, and care for burn damage. Cutaneous treatments use clays with a strong capacity of ionic exchange.

[1] Two complementary pieces of information: the dimension of clay particles may be reduced by crushing during preparation: thus clay can become dangerous because of the presence of very small crystallites. The same type of clay can contain different compensation ions and be better adapted to certain uses.

[2] The object of pedology is the study of soils.

Heat storage capacity is another important parameter. Heat contained in a paste causes the perception of a softening effect, which has an unquestionable psychological impact and acts on the results. Numerous physical and physicochemical properties take part in this function (Yvon and Ferrand 1996).

Natural clays are often associated with other finely divided minerals, and it is this mixture that is put in contact with water. A traditional method of preparation consists in putting clay powders in a great quantity of water; clay is then in suspension in water. This suspension or dispersion named *clay milk* is often used as a drink but it can also be used for body ablutions. There is a strong similarity between this liquid and thermal spring water loaded with elements present in the geological layers through which they have made their way (on this point the book by Pomerol and Ricour, 1992, may be consulted). The elements present can be released in the form of ions or dissolved molecules (they are then in a degraded form), or in the form of particles (crystalline or paracrystallines) in suspension.

In drug manufacture or products intended for body comfort and health, clays are often used particularly for their adsorption surface properties which make it possible to fix active chemical groups. The slowly developing selective release of these groups allows their use with a "delay effect." Clay minerals' surfaces constitute a simple vector of the therapeutic molecules used. The duration and speed of retention or separation between active group and mineral support can often be controlled; one thus obtains delayed actions, very useful within the framework of certain drug therapies. Numerous molecules such as vitamins, antibiotics, enzymes, or hormones can be adapted to this technique of fixing by adsorption or absorption on the surface of the clay mineral layers.

The same property of fixing on the surface of the layers can be used to trap or eliminate undesirable molecules. Some deodorant actions or purifying clays are well known, such as collecting those molecules responsible for nauseating odors from some wounds, fixing of tar of tobacco smoke, and deodorization of litters for small pets. Clays can fix a large variety of molecules, some of which are toxic: they prove to be very effective natural filters. It is advisable, however, to be vigilant that the applications do not have an indirect and undesired harmful action, in particular on the respiratory system (mineral atmospheric dust fixed first by the fur of the cats using the litter, e.g.).

Massage techniques practiced with muds, gels, or clays in suspensions in liquids also exploit their light abrasive effect (gumming and exfoliation). Beauty care also uses this property for shampoos and soaps. Manufacturers of toothpaste containing clays optimize the effectiveness of their product by possibly combining the light abrasive effect and the properties of absorption or exchange.

Dealing with relationships to health, clays cannot be neglected in the food field. They are mainly present in the "grounds or earth" used in geophagia, and this animal and human tradition dates back to antiquity. Sometimes they are used as mineral fillers, to manufacture gels or suspensions, or to purify liquid foods (water, oils, wines, etc.). They are also used very effectively in cooking to preserve food from contact with air and by coating food with clay to ensure tender cooking.

The clays transformed into ceramics also take part in the make-up of kitchen utensils.

Clays are often close to daily life and consequently the benefits that can be expected from them seem to be well known. But, although they are really efficient, traditions often convey inaccurate information. Consequently we deal with concepts aiming at improving this knowledge. For that, it is necessary to learn how to distinguish pathologies themselves and the properties from clays associated with their treatments. Comparison of these two concepts makes it possible to determine a more efficient therapeutic action. This field is very vast and sometimes debatable. Nevertheless, what is unquestionable is the effect felt by patients, whether real or psychological (because even in this case, this effect can reinforce the action of other therapeutic acts). Perception by the individual of a pain, or a sensation of discomfort, is often difficult to connect to its pathological cause. The result of this may produce confusion between causes and effects, and this has a role in the use and the result of the traditional uses of clays. All this can be complicated by how the clay is implemented.

Let us recall some uses among the most current:

1. *External uses*:
   - Envelopments, cataplasms or poultices, compresses, realized with a clay paste
   - Clay milk or clay water, mud baths
   - Cosmetics
2. *Internal uses*:
   - Gargle and mouth baths
   - Digestive tracts
   - Rectal injections
3. *Other uses*:
   - Food, kitchen
   - Purification (water, other liquids)

Appendix C shows the potential applications in the medical field of clay and clay minerals that have been reported and surveyed in the literature; it is advisable on this point to be careful with respect to the stated possibilities because the majority of them do not have scientific support. However, some have already received confirmation by scientific studies or private clinics; they are then recognized as effective and can be validated. Others, less relevant or also less financially profitable, sometimes remain incorrect in the way they are presented or justified. At times, it is even necessary to give up the application because these assertions are simply groundless. Various proposals of this list and various methods of classification must be planned in order to join each type of need to a type of real physicochemical clay property and justify the choice of the most suitable mineral. In this way, a simple vision is set up which makes it possible to eliminate inaccuracies that often accompany the therapeutic use of clays in the absence of scientific data.

We have to specify here that in all the current applications, even if they are qualified as "internal," the use of clays is not necessarily made inside a living organism itself. It implies a direct physical contact of accessible zones on the exterior of the body in natural ways (skin and mucous membranes of the digestive system). The mineral itself always remains external to the treated tissue. Moreover, in the actual position of medical practices and knowledge, lungs and their respiratory tracts for which the fine clay particles are comparable to generally harmful (even sometimes fatal) mineral dust are completely excluded from our study. Hearing and ocular methods are also excluded from these applications.

However, it is physically possible during the applications considered here that very small clay particles, groups of atoms, certain atoms, or ions can diffuse indirectly in tissues and even reach the mucous membranes of the respiratory tract. This situation will need special attention in the case of very long and intense, or abusive or unconscious treatments (work or places of residence where the atmosphere is contaminated). The most severe contraindication relates to the use of fibrous phyllosilicates, such as chrysotile which is indisputably responsible for serious pulmonary mesothelioma. These heavy pathologies, rather close to those better known of silicosis, imply that these minerals are inhaled. Their ingestion must be strictly limited, and in any case reduced to a strict medical indication. The same safety-first principle and of preventive measure also requires being careful with two other fibrous clay minerals, sepiolite and palygorskite, whose applications are appreciated, but which remain potentially dangerous.

6.2 Complexity of the Interactions

If planning to focus attention on the mechanisms concerned when using clay intended to improve the health condition or comfort of an organism, the following course of action should be observed.

1. Those of physical or energy types such as the diffusion of heat through the contact surface
2. Those of chemical types bringing into play transfers of matter of one medium to the other (e.g., by osmosis)
3. Those that act via the consciousness of the patient and are essential for him on a psychological level (e.g., concept of well-being or that of relief compared to a pain)

Those applications having usually recognized favorable effects can only validate one positive interaction between the mineral medium of clays and that of living organisms. However, bringing these two apparently very distinct and opposite kingdoms of matter closer is not obvious. Without wanting to justify a priori a cause-and-effect relationship between the properties of clays and their positive applications to living organisms, one can evoke some comparative data that can contribute to an understanding on the key aspects of this interaction.

In spite of their differences, there are many common or convergent points that make it possible to look in a similar way at certain reactivities of living organisms and clay minerals.

In terms of differences, it can be seen that living organisms and clays have very different chemical compositions and structures: the living one is primarily made up on a carbonaceous basis, whereas the clay minerals have a siliceous basis. Thus the connections between the chemical elements are primarily covalent for the organic molecules of the living world whereas they are especially ionic for clay minerals.[3] A living organism *is formed by a community of molecules*, extraordinarily complex and organized in cells communicating between them by chemical signals. In the mineral world, a clay particle is formed by many individual layers more or less strongly dependent between them because of physicochemical laws. The basic entity is described by an elementary mesh that contains a crystalline unit cell the size of a few Angstroms. (Note: the nanometer must now be adopted.) This unit cell is repeated to form a rather simple crystal, of nanometric size, and whose defects are at the origin of the principal physicochemical properties.

In terms of similarities, one can recall the following aspects.

1. Just as a mineral microcrystal has, a living cell has an external surface characterized by specific properties that ensure part of their relations with the external medium
2. Atoms constituting these matters have their own reactivity
3. The magnitude of the implicated bonding strengths is nearly similar
4. The two types of matter contain water, the molecule fundamental to life. A living organism, whether human or animal is composed of water for 60% of its weight. Clay minerals,[4] under the conditions of natural ambient conditions, also contain an important percentage of water of about 20% or more and variable according to the species, but never zero
5. The reactions of adaptation compared to the modification of the physical conditions of temperature and pressure are not very far apart, in particular regarding the action on water (hydration, dehydration)
6. Clays and living organisms have an important activity under normal conditions of temperature and pressure
7. The interactions between clays and living organisms take place at the interface between these two media, the results measured by the diffusion of atoms, ions, or molecules and very often implying a liquid phase and water molecules

[3] In a connection between two ions of opposed charges, the negative ion brings an electron to supplement the external layer of the positive ion. In a covalent bond, generally between atoms of similar electronegativity, each dependent atom shares an electron of one of its external layers to form a doublet of common electrons binding the two atoms.

[4] The quantity of dependent water contained in a clay is measured on a sample whose initial mass is stabilized at 20°C before any rise in temperature. For certain very absorbing clays, it is necessary to specify the pressure and the ambient water content.

In terms of assessment of the relations with ambient conditions, their behaviors are definitely different while remaining comparable. A living organism always tries to limit a disturbance and avoid its destruction; it "fights" for its existence, its survival, either by reflex or consciously. The cells have an aptitude to be mobilized collectively and hierarchically to control an intruding element such as a microbe or to repair a broken bone. For that, certain cells can reproduce themselves. In the case of a clay, its survival or its safeguarding in time (passive in this case, but nevertheless satisfying the laws of balance and stability of the system) imply a structural evolution giving access to a more stable energy state. In this evolution, clay platelets most often divide themselves parallel to the shape of the elementary layers producing increasingly thinner and thinner platelets or crystals.

The concepts of consciousness and memory deserve special attention because it is possible to confuse them. Although being very comparable, at the same time they make it possible to separate and bring closer the two worlds considered. Living organisms and clays are reactive media able to evolve and adapt or conform themselves to a new environment. Consciousness influences the reactivity of the living. The chemical and structural memory of the clay minerals, which depends on their conditions of formation, is a determining parameter that does not imply the concept of will. It is the energy of the system and the physicochemical conditions that impose their laws and condition their successive evolutions.

We thus very often note that a mechanism acts on another and that there is a priority in the succession. To recognize a mechanism, it is necessary to be able to distinguish the cause from the effect clearly, both being able to be dependent by a concept of reversibility. This delicate point is at the origin of the majority of bad interpretations and hasty assertions met in daily life in connection with the action of clays.

Some phenomena, such as the perception of heat, can be perceived as something pleasant or definitely unpleasant. Their perception is not really definable; in any case it is hard to quantify. These are well-known facts that should not be neglected in the total analysis of the product. It is also necessary to be wary of many other aspects, such as that temporary relief can make it possible for other mechanisms to exert a repairing role, a relief expected by the patient, but entailing a possible confusion of causes.

6.3 Reactivity and Mode of Clay Action

Before ruling on a mechanism of action of a specific application and selecting clay for the organism that profits from the application, it is important to know the reactive potential of clays in the light of current knowledge.

6.3.1 Various Types of Reactive Surfaces Related to Structure

It is a rule that the smaller the dimensions an object has, the more important its surface becomes in relation to its volume.[5] Microparticles of clay minerals have the characteristic of being formed by the stacking of very fine and of very small size layers, each with two plane basal surfaces. The external surface of the particle includes an upper surface, a lower surface, and side surfaces. All these surfaces are particularly active in fixing ions and/or molecules more or less strongly.

Figure 6.1 schematizes these concepts for a particle consisting of a single layer (Fig. 6.1a) or for a particle consisting of a stack of several layers (Fig. 6.1b). Dimensions of elementary layers[6] piled up in parallel one upon the other are about a micrometer in the direction of the plan of the layers, but their thickness is thinner, about some nanometers (these dimensional ratios are not respected in the drawing).

Each of these layers is made up of very simple structural elements whose pattern is valid for all the phyllosilicates. Initially let us consider the layer of kaolinite. It is constituted by the stacking of two well defined layers, composed of ions; one is formed by tetrahedrons centered on an ion Si^{4+} and the second one by the octahedral centered on an ion Al^{3+}. Kaolinite belongs to the family of phyllosilicates the layers of which are composed of a tetrahedral layer and an octahedral layer, named TeOc for short (Figs. 6.2 and 6.6).

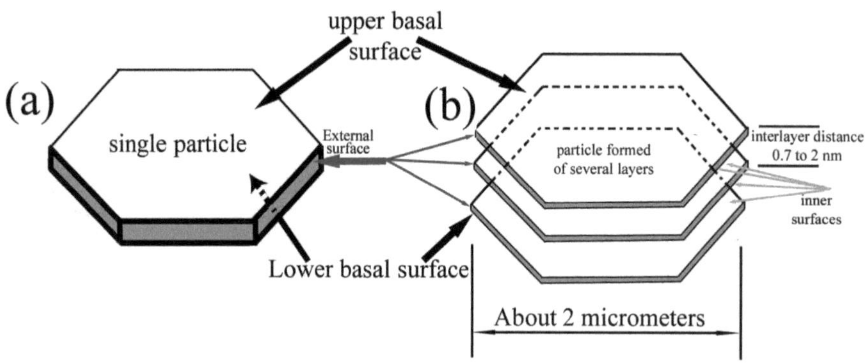

Fig. 6.1 Various types of particle surfaces; (**a**) for a single layer particle (on the *left*): one upper basal surface, one lower basal surface and side or edge surface (made up of several facets); (**b**) for a particle constituted by *N* layers (on the right): one upper basal surface, one lower basal surface (2*N*−2), inner surfaces, and *N* side or edge surfaces

[5] Let us take the case of a cube on side **a**: its volume is $\mathbf{a}^3$ and its surface is $\mathbf{6a}^2$. The relationship between surface and volume is equal to **6/a**. This ratio takes a high value as "**a**" is smaller. See Appendix D.

[6] The thickness of the layers, by including free space between two layers, is 0.7, 1 or 1.4 nm, according to the type of clay; refer to Sect. 5.1.1.

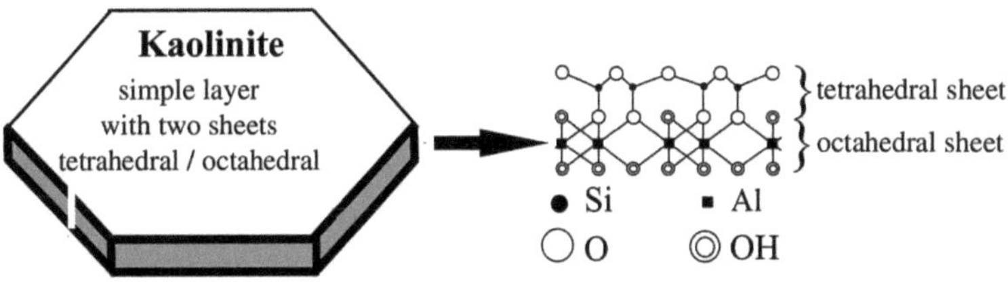

Fig. 6.2 Composition of kaolinite layer, TeOc type

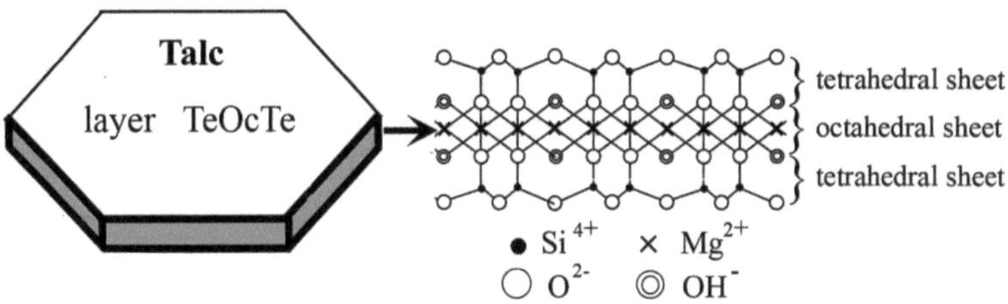

Fig. 6.3 Composition of talc layer, TeOcTe type

The second case we must consider is that of talc whose layer is constituted by stacking three layers, one made of octahedrons centered on an ion Mg^{2+} and surrounded by two identical layers formed by tetrahedrons centered on an ion Si^{4+}. Talc belongs to the phyllosilicate family whose layers are named TeOcTe for short (Fig. 6.3).

Because of their very small size and structure, clay particles have a very reactive large surface compared to their volume. As an order of the value of reactive clay surfaces we note that in the case of a smectite (layers of TeOcTe type), the maximum surface accessible for 1 g from minerals reaches the enormous value of 845 m^2. This figure alone clearly justifies the importance of the surface properties regarding accessibility to the two basal faces of each layer of most TeOcTe clays.

This is not the case of kaolinite, quoted above, from which the internal basal faces of the layers are not very accessible because there are no compensation ions; consequently it remains in the shape of particles made up of several layers, which decreases the reactive surface.

6.3.2 Various Reaction Scales

At this stage of our explanation we must note that the main part of the reactive properties of clays is in relation to their surfaces. This basic concept of surface reactivity depending on the electric charge must be approached by several steps:

1. The elementary microcrystalline particle
2. Interactions between particles

3. The association of a very large number of particles forming a paste
4. Interactions within a complex medium (gastric, e.g.)

The great difficulty in interpreting and exploiting the properties of clays is understanding the effective scale of the considered mechanisms: nanoscopic (some nanometers), microscopic (dimension of the micrometer), macroscopic (of the millimeter to several centimeters), or mesoscopic.[7] Scientific work on these changes of scale is far from being complete: physicists and mathematicians endeavor to come up with models that make it possible gradually to get a good explanation of the properties at the passage between each dimensional scale.

When a large quantity of clay microparticles is rearranged in a dispersing medium, there remains between them an interparticle porosity where various materials can be inserted that are then applied at the same time as the clay mineral. In addition, the very small dimensions of the pores are at the origin of the physical properties that supplement those of the mineral itself.

Figure 6.4 schematizes some fittings of layers and particles, in weak concentration, in water and the various associated types of porosity.

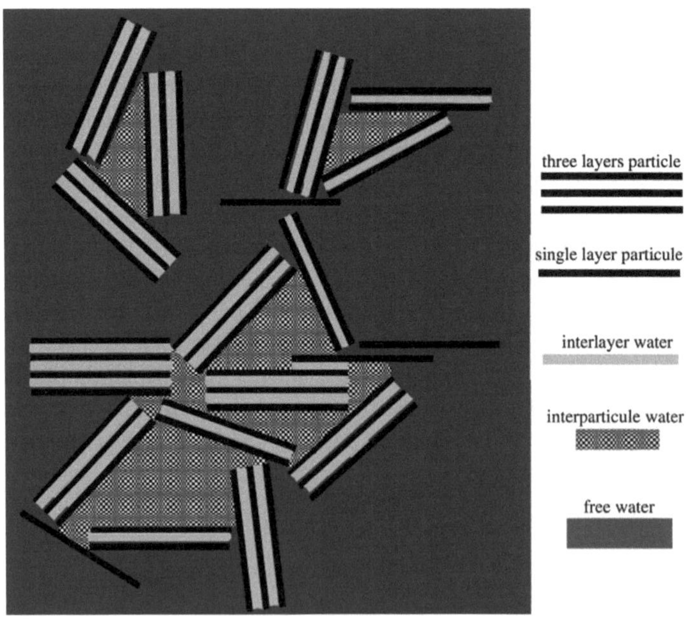

Fig. 6.4 Schematic representation of the various types of porosity. Observed in a plan of a section perpendicular to the plan of the layers

[7]Mesoscopic: Belonging to an order of magnitude located between the limits of the macroscopic world and that of atoms and molecules; on a scale where the structural description of the matter presents characteristics and properties which concern both quantic and traditional physics.

A clay particle is a crystallized solid body that occupies a well-delimited volume in space, very small but not negligible. This microscopic volume is structured in layers of great side extension and low thickness. Inside the mineral structure, the components are subjected to the crystallographic order and ionic connections. On the level of the external surface of the particle, the relations controlling the cohesion forces between the ions of the mineral skeleton itself become nonexistent. This field constitutes a major accident for the crystal which does not have any existence beyond this limit. Its result is a strong energy imbalance in the vicinity of the external surface. The material is unable to be satisfied with this situation; the clay particles are naturally supplemented by hydroxyl groups (OH^-) or by protons (H^+) ensuring a temporarily stable situation.[8] Starting from this surface and with a growing distance compared to the particle and beyond, other types of connections are created with ions, atoms, or molecules in the vicinity; they allow the return in a more stable energy state, better adapted to the nature of the medium. This state is only transitory: at any moment it is likely to be modified according to a variation of the composition of the medium where the particle is located.

Taking into account the low dimension of the particles, one can compare clays to materials with colloidal behavior (the analysis of the mechanisms is complex; readers can refer to the publication of Van Damme 1994). Space, time, and energy balance are three nondissociable parameters of the life and state of clay.

Note: It is important to understand clearly that the reactivity of clay is based primarily on mechanisms of a physicochemical type which find their origin in the larger side extension of the elementary layers and in the existence of the structural defects. For clay to be active it is necessary that the selected mineral, the medium in which it is dispersed, and the organism to which it is applied, can interact. In other words, so that the reactions are possible, and carried out, it is necessary that each of the two media (clay mineral and living organism) should be prepared to give maximum accessibility between the active sites of the mineral and that of the organism.

6.3.3 Various Modes of Action

For a therapeutic use, we consider clay in its largest sense which includes the mineral particles themselves as well as the elements or associated molecules and are fixed on the clay particles thanks to their surface properties or are dependent on the continuous medium used for the selected application (powder, paste, mud, or suspension). According to the case, ions, atoms, or molecules unrelated to clay but present in this medium also take part in the reactions and contribute to the effectiveness of the application. We approach the origin of these various elements by initially describing in this section the process of mobilization of the material permitted by the clay structure. It should be noted, however, that the opposite mechanism is also possible thanks to fixing and compensation possibilities.

---

[8] It is obvious that these two ions are not the only ones able to intervene; nevertheless they are most common.

6.3.3.1 Mobilization and Transfer of Elements from Clays

The structural characteristics of clay minerals that are advanced briefly here were presented in more detail in Chapter 5.

Let us initially consider the elementary clay layer, that is, the single layer (with two or three sheets), and the associated compensation ions. The electric surface charge of the layer is stabilized primarily by the fixing of ions, ions hydroxyls or protons, whose connections can easily be broken. The proximity of an active group that develops an electric potential around it able to ensure a delocalization of electrons or an electronic transfer is enough to break such connections. The acid or basic character of the medium or the presence of polar molecules (such as water or alcohol) determines the type of physicochemical reaction. Two mechanisms are then possible: fixing or exchanging, primarily on the higher and lower faces of the layers, but it is also advisable not to neglect the side faces of the layers.

If one considers a microparticle formed of several layers, interlayer elements can be mobilized and moved between these layers. Their diffusion is slower than on the level of the external surface, but it is carried out all the same with the assistance of the hydration water molecules of the compensation interlayer ions, and the mechanism of swelling.

Finally if one considers a group of particles dispersed in a fluid, the element, whose connections with the clay support are broken, becomes mobile in interparticle porosity. Its circulation is then possible and its displacement occurs in the direction of the zones of weaker concentration if the clay medium is almost in contact with the skin of the patient (or mucus re-covering the intestinal membrane); gradually the clay medium undergoes a modification on a low thickness and a gradient of concentration is established. It allows the continuation of the process of diffusion, or on the contrary, its arrest.

6.3.3.2 Fixing and Trapping of Elements on Clays

The surface of clays has properties of adherence and fixation that can intervene in extremely varied ways:

1. The particle can adhere on a surface and be fixed on it (skin, intestinal surface, etc.)
2. The particle itself can fix different and more or less complex entities (ions, molecules, very small particles, etc.)

When the clay layers are fixed on a support, the adherence that results is effective and the concerned forces can be of various types:

a. Simple Van Der Walls type attractions
b. Action of the compensation ions acting as centers of attraction
c. Electric charge of the side faces of the layers that allow fixing on the sides of the particles

These properties of adherence and fixing make it possible for clay to cover an uninterrupted exposed surface while acting as a protective bandage, or more or less completely enrobe the small ailing element to insulate it from the harmful organism. The target can simultaneously be a continuous surface such as the skin and small more or less mobile objects such as intestinal microorganisms or toxic bacteria.

From their structures, clays present a certain proportion of external surfaces corresponding to the border or edge faces of the layers. For the majority, the edge surface (fixed by the thickness of the layers) has a low value compared to that of basal surfaces of the layers. However, there are particular clays, such as the fibrous species sepiolite (see Fig. 6.5) and palygorskite, that are made of pseudo-layers; they have a very important proportion of side surfaces compared to basal surfaces: two thirds for the sepiolite and one half for the palygorskite.

Figure 6.5 shows the structural channels, the importance of the side faces or edges of the layers, and the various accessible sites. Under normal conditions these channels contain water.

We voluntarily gave examples of the basic idealized models that are very close to reality. Each case is particular and must be studied specifically. It should be remembered that the elementary layers are not all independent and individualized (the extreme case we have just treated above). They are gathered in "packages of layers" again called "particles" in our text, bound between them by attractions due primarily to charge imbalances. The number of layers superimposed in a particle is characteristic of the origin of the clay and constitutes an important parameter for the examination of its properties.[9] These particles themselves are gathered in aggre-

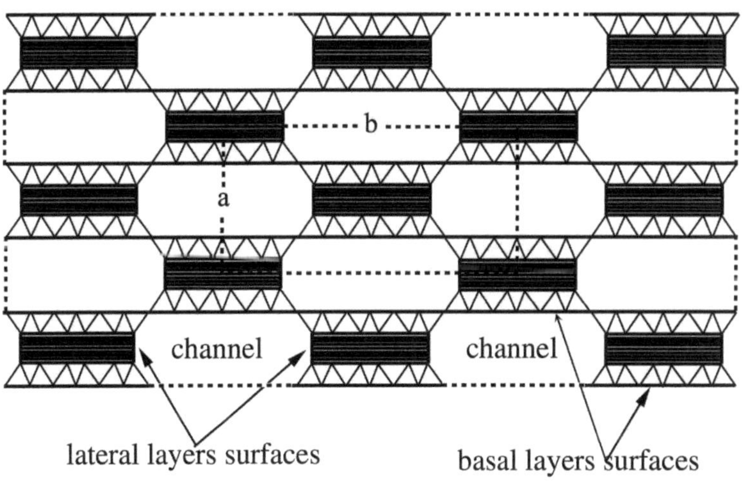

Fig. 6.5 Diagram of the section perpendicular to the axis of lengthening of a sepiolite fiber

[9] The number of associated layers to form a particle answers a statistical law, suitable for the sample, known as the function of distribution of the number of layers. This law is in close relationship to the conditions of genesis of the mineral kaolinite.

gates. Using high-resolution electronic microscopy we show some stereotypical examples of the natural stacking of layers within a particle of kaolinite (Fig. 6.6) and a smectite contained in a bentonite[10] (Fig. 6.7). These two stereotypical pictures were obtained on ultramicrotomy cuts of 50 nm thickness of hardened clay in a resin, carried out perpendicular to the direction of the plans of the layers, which thus makes it possible to visualize the layers in the form of parallel lines.

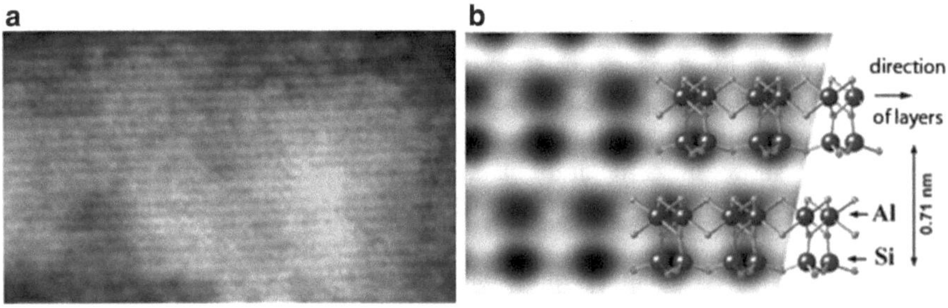

Fig. 6.6 Visualization by high-resolution electronic transmission microscopy (TEM) of kaolinite structural layers; (**a**) the structural layers of kaolinite from Charentes (one distinguishes the kaolin that represents the extracted raw rock and kaolinite, the main clay mineral that constitutes this rock) are plane and rigid and the interlayer distance was estimated at 0.7 nm, 1990; (**b**) very high-resolution electronic microscopy shows the octahedral and tetrahedral sites, 2008

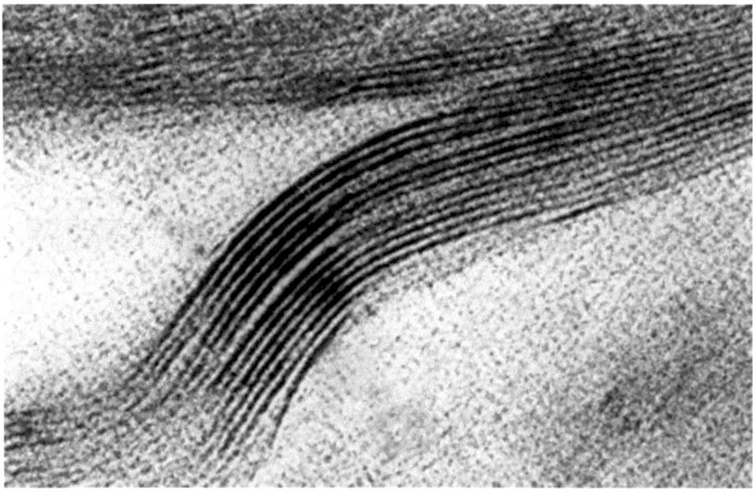

Fig. 6.7 Smectite: Montmorillonite of Wyoming; transmission electron microscopy image of montmorillonite shows that the structural layers are flexible, and that the interlayer distance was estimated at 1 nm

[10] Bentonite is a rock (and not clay) that mainly contains clay minerals of smectite type.

6.4 Specificities of the Modes of Application

6.4.1 *Complexity of the Relations "Need–Application–Effect"*

Specific papers mention many applications of clays in the field of health or well-being. The list of pathologies, given in Appendix C, is not exhaustive and can still be supplemented, because some reviews, works, or practices concerned with oral traditions can have escaped us. It is scientifically very difficult to accept such an inventory from the start, which led us to adopt an alphabetical classification rather than a thematic one.

First of all we consider that the applications must be close to a property (or a type of properties) of the clay mineral. The relation between the precise health need and well-defined clay is not unique. Indeed, vis-à-vis the demand it is initially necessary to consider a "family" of clay rather than a "type" of clay, and among the members of this family to specify the character adapted to the need. From another point of view, all organisms are different and cannot react in the same way to the same type of clay mineral. We specify this step using examples:

1. To carry out a body gumming and exfoliation, it is necessary to have clay whose particles are rigid, which excludes the smectites but preferentially uses illites or kaolinite.
2. To manufacture shampoos the effects of gumming and absorption are necessary but cannot always be used simultaneously. In this case a mixture of two clays, for example, kaolinite and smectite can be used.
3. To facilitate the softness of a dry contact, the sliding effect of the layers or particles over each other is sought, an important property of talc.
4. To make a stomachic dressing, the different roles of absorption and adsorption are essential; smectites are then appropriate.
5. For people with metal prostheses this point must be taken into account because the ionic exchanges can have an action on these materials.

Before choosing an application it is necessary systematically to carry out a precise analysis of the needs of the organism that will be treated, but this analysis is not enough by itself: it is also necessary to associate knowledge of the properties of available clay minerals. Then, keeping in mind the results obtained is obviously very important. Updates of these results will eventually determine the change of certain treatments that one could develop but which appeared dangerous. There was the dramatic case when molecules were added to talc (classically used for the care of babies and that by itself is not harmful at all); these molecules used as disinfectants diffused like poison in the blood of the babies.

Clay has only exceptionally a direct action. In general, there is an intermediate factor that intervenes between clay and the object to be treated. It can be an ion, an hydroxyl, water, a molecule, or a complex, generally placed in an interlayer situation. All kinds of situations are imaginable; they make it possible to foresee a very large field of applications.

Sometimes clay already contains in itself the useful element whose action is wished. For example, if the compensation ion sodium is already present in interlayer space, it can be exchanged because the requirements are carried out. When this exchangeable element is not there or not appropriate, it should voluntarily be introduced during the preparation of the clay before being used.

Clay is neither easily nor quickly permeable with a flow of molecules. Its preparation requires long and often complicated processes, which are reflected in the marketing price. We must once again point out a concept that has become current: the "time of maceration." It gives access to "the suitable state of maturation of the mixture containing the clay" for which one hopes that the reactions of fixing are favorable to the envisaged applications. Many physical methods make it possible to ensure the validity of a preparation. In particular, the position, the nature, and the force of fixing an agent can be located to ascertain the structure of the layers of clay. Unfortunately, too often it is not possible financially to assume such research. One thus sticks to an empirical solution, often good, but not giving any pertinent certainty of the origin of the properties.

Many factors intervene in the maturation of the preparation:

1. The water quality employed that presents a charge in dissolved chemical elements (often in the ionic state), in organic compounds and microorganisms, and has a given pH
2. The concerned clay, whose nature and composition condition the physicochemical properties
3. The maturation time and the maturation method being employed (with agitation and/or rest)

This stage brings unquestionably important modifications to the nature of the involved elements. The crystallochemical properties of clay evolve until a new state of balance and *in fine* confer to it notably modified aptitudes of exchange and transfer. Because of its complexity this objective is always uncontrolled.

The various elements present initially in the mixture constituting the peloid can be assembled very differently and thus constitute undesirable compounds or have undesired physical properties (residual radioactivity, e.g.). These elements can travel through the organism by the mechanisms of diffusion that we discussed. Very great care must be taken in the course of this process; it must be accompanied by suitable analyses imposed on the establishments of care. This excludes the individual use of these applications.

In the fields of health or well-being, clay experts must permanently consider various approaches that we have already presented. At this stage our scientific role is to connect empirical aspects between them (which are always associated with expressions of relief or satisfaction, therefore subject to human perception) and proposals for scientific models based on always improved and renewed knowledge.

Clays can act in two ways:

1. As a physical interface or a mechanical barrier
2. As a chemical interface and an exchange site

These two modes of action can be simultaneous, but with distinct kinetics:

1. An immediate effect is associated with direct physical contact and a mechanical or thermal action: for example, the case of talc when used dry, whose electrically neutral layers slide over each other without friction, which confers real and incomparable softness to it.
2. A slower effect is associated with a chemical action implying, by adsorption or absorption, the properties of exchange and fixation ions or molecules that can then be released, fixed, or exchanged for medical care. This effect is valid for the adsorption of microorganisms by particles.

In the natural layers, geological conditions can subject the particles' clays to a strong pressure compacting them in very resistant blocks when they are dry. To benefit from the presence of the many elementary particles constituting this compact mass, it is initially necessary to separate and individualize them (this is the role of the preparation process and packaging of marketed clays; see Chapter IX). But these minerals also exist in nature in the form of pastes, muds, and suspensions in water, and natural reserves can then be used directly in the field of interest. A preparation is generally necessary; it can be accompanied by purification, an operation concerning the elimination of nonclay compounds (such as sands, oxides, or other impurities) that are often present in the natural products. Organic compounds can accompany clays but more rarely.

Purified clay is then more or less strongly diluted in water according to the envisaged use and objectives to be reached. The various presentations of the clay are thus differentiated: dry powder, paste, or suspension according to the importance of dilution or more exactly the rate of hydration. These three states must be defined to exploit their possibilities to the maximum. Although these empirical concepts are generally well understood, it is desirable to refine our knowledge. In fact, the affinity of clay for water is so strong that any variation of the water content acts considerably on its physicochemical properties.

In the space where the interaction must be carried out, the mineral phase and the associated liquid phase have variable relative proportions defining that which is included in the other. Moreover, the clay phase has an aptitude for the swelling that can result in certain cases in justifying a strong action on the associated medium (dispersion of the latter, installation of a gel compression with the contact, inversion of the concept thinned/diluted).

We must recall that the majority of uses of clays in medicine and alternative medicine always imply a direct contact between clay and the organism. This zone of contact corresponds to an interface between two distinctly different media, one mineral and the other biological. Obviously, the organism itself can also be prepared to obtain better effectiveness. Different exchanges occur through this surface:

1. Those of physical nature: transfer of heat, transfer of the forces of pressure and tension
2. Those of chemical nature: exchanges of ions or molecules between the two media

Certain chemical exchanges correspond to the phenomenon of dialysis or osmosis through a permeable membrane, such as skin. The exchanged ions or molecules can be naturally present in the water–clay system or be artificially fixed beforehand on or between the clay layers (by the mechanisms described further down), or be mixed with the substrate used to constitute the paste or the suspension.

These considerations should be omnipresent at the time of the use of clays.

6.4.2 Fundamental Knowledge for Users

We have just described the simplest elementary applications where clay is in direct contact with a well-defined target and intervenes with a reduced number of parameters. A simple example of a situation is the direct application of clay on skin.

Often reality is much more complicated, in particular in the case of a use that brings into play the digestive function, because in this case, introduced clay is mixed with an alimentary bolus which moves from the mouth to the anus, and this extremely complex situation makes the reaction mechanisms very delicate to interpret. Another complex situation is to be found in beauty care: it results from the addition of many molecules to the clay (necessary for obtaining the properties wished in this field). The interactions with minerals are then not always well known or easily controllable.

Facing such complexity it is advisable neither to give up nor to leave routine and empiricism to deal with the situation. This complexity compels us to return to basic analytical concepts. As the description of this medium is difficult when considered in its entirety, the concept of scale becomes of major importance. Some of all the factors present simultaneously can be described on a dimensional scale adapted to them. It is the case of clay particles whose crystallographic size and characteristics are known and consequently its average nature of the reaction interface. It is then advisable to analyze the phenomenon considering the adapted scale to specific dimensions of these particles. Gradually, it is then possible to define a "field of coherence."[11] This basic concept makes it possible to take into account, using the local scales thus defined, a zone where the reaction properties of clay are known and can be correlated with those of the medium in contact with them. Beyond this zone of coherence, another field of coherence is essential and the properties can vary according to this new local context.

6.4.3 Time of Application

We often used the terms *transfer*, *exchange*, *diffusion*, or *fixation* which all imply the time parameter. Indeed the length of time has great importance in the use of clays with therapeutic purposes.

[11] The "field of coherence" defines the zone inside which the parameters controlling the properties remain the same. It is thus necessary to add a qualifier to define a field of coherence. This field of coherence is also where a homogeneous reaction mechanism prevails; beyond this limit it can vary (e.g., if the local composition in exchangeable ions is different).

This concept is easy to understand in the case of thermal evolution. Heat moves in the direction of the coldest material, and the speed of diffusion is dependent on the quantity of heat and of the thermal conductivity. It is sometimes more difficult to define the length of optimal action during matter diffusion, in the form of ions or molecules, for example. This is why we point out a basic principle that governs these phenomena: it is the law of mass action[12] which stipulates, *if possible, that an unbalanced system evolves so that the higher concentration decreases in order to approach a state in steady balance*. Thus if the organism in contact with clay can absorb ions, these ions diffuse through the skin in order to decrease their difference in concentration on both sides. In general there is an exchange of ions, sometimes of molecules. Displacement can be done in two directions: increase or extraction.

In our case the mass of the treated organism, as well as the quantity of clay used, is well known. The evolution during the time of application can then lead to a balance; thus it is unnecessary to prolong the action indefinitely. For example, there is no more action when the chemical concentration is the same in the organism and the clay mass with respect to a determined element. It is identical for heat. That leads us to renew clay material when the use implies a chemical compound whose concentration is becoming exhausted by transfer in tissues, meaning its potential of action is exhausted. Obviously when it is a question of insufficient heat, it is always possible to warm up the clay mixture, by simultaneously carefully supervising the state of hydration.

The estimate of application time is controlled by the practitioner. For the patient, this duration often implies a very different psychological concept and brings perception into play, which is initially that of relief, of well-being brought about during medical care but also that of tiredness generated during those periods of care (all users report this simultaneous sense of relief and tiredness). This relaxation is also a factor facilitating the action of the selected therapeutic method but this is a question beyond the field of exact sciences, even if the interaction between these two aspects is in fact very important and cannot be neglected in any case.

6.4.4 Determining Stages for the Optimization of Clay Effects

We expressed in the foreword that the various clay mineral species are comparable with "*structures in constant evolution*" whose properties vary according to the geological context of the deposit (natural source) and in the way in which the material is exploited (extraction and application). We must thus distinguish various periods or stages during which big changes take part in the setting of the characteristic properties for the use of clays:

---

[12]The law of mass action determines the balance of systems in contact with one another and between which an exchange of matter is possible.

1. *The stage of investigating the origin, localization, and nature of clay*: The local geological history and the context of formation of the deposit (age, localization, nature of the surrounding rocks, climate) control the nature (crystalline structure, chemical composition, relative abundance) of the clay mineral species present and their potential of physicochemical reactivity.
2. *The stage of exploitation of the deposit and packaging of the product* before any decision of the type of use: Some of these operations may not be quite harmless (purification, crushing, etc.).
3. *The stage of preparation with a well-defined aim*: Edible clay ingestion to fight gastro-intestinal ailments, or peloid/mud to be used in balneotherapy to treat, for instance, osteoarthritis and other muscular–skeletal ailments. Their preparation sometimes implies some complex mixtures with liquids, active compounds (ionic, polar, etc.), and of neutral or living bio-organic compounds. In addition there are physical temperature variations and a period known as adapted "maturation," justified by the long experience of practitioners. Much can be said to describe the complex evolution of the mixture during the time of maturation. There are very many parameters that intervene and their analysis is complex. However, scientific experimenters manage gradually to obtain some important data from their experiments (Veniale 1996; Veniale et al. 2004; Droy-Lefaix and Tateo 2006). The preparation can usually take several months to obtain changes and a new balance. Despite everything, practice remains the leading guide to determine the moment when the product is ready. However, some quantified indicators gradually come to help the decision, for example, measurements of viscosity, elasticity, or measurements of affinity by adherence to the receiving media. These measurements are useful indicators but they still require many studies to become applicable in all cases and even in everyday use.
4. *The stage of marketing*: With packaging operations, storage, and transport.
5. *The stage of preparation preliminary to the application:* It must be carried out with the greatest care. Contamination must be avoided at any cost (water or liquid employed, containers, mixing tools, application spatula, etc.). The nature of materials constituting the container and the tool being used to carry out the mixture are often evoked. It is often recommended not to use metallic recipients or tools; this could be connected to the ionic constitution of clay minerals but actually nothing is really proved. It is better if possible preferentially not to use metal objects.
6. *The methodology of mixing:* Particular attention must be paid here and especially to the exactitude of the quantities of clay and water present; the physical status of the mixture is indeed very sensitive to the rate of hydration. The rest period of the preparation is also important data.
7. *The stage of application on the organism:* The simultaneous coordination of the receiver (skin or mucous membrane) and of the clay mixture can be ranked with a new stage of maturation, where the organism intervenes to accept or provide ions. This application time is often brief compared to that of the maturation of the peloid. However, we cannot basically distinguish these two stages of reorganization and evolution of the media.

6.5 Principal Mechanisms of Action and Application

According to what has just been presented, clay intervenes in mechanisms of action that can be divided into some general groups:

1. Barrier effect more or less impervious
2. Import of matter by clay
3. Exchange matter between clay and organism
4. Matter fixed on clay and its elimination
5. Transport of matter between clay particles
6. Physical action: heat transfer or mechanical (e.g., pressure)

Note: "Matter," depending on the specific case, could be molecules, atoms, or more or less complex ions that have, by their composition, a reactive affinity with clay; it can also be microorganisms.

The modes of action suggested here compel us to recognize the fundamental existence of a physical border that separates the organism's interior and exterior and constitutes the skin or the mucous membranes. This barrier allows transfers of energy (heat, pressure) or of matter (water in the example of hydration/dehydration) in both directions. The preparation of a curative method employing clay implies taking into account these data and assessment of the probability of their effectiveness. In this field of traditional methods (presumably soft, but unfortunately sometimes harmful), special attention must be brought to distinguish perceived symptoms, causes, and effects. The application of clay can often reciprocally act on one or on the other, independently or simultaneously. A true analysis of the possible effects must be made, which imposes in this case recourse to the only possible and accurate interlocutor, the physician. The latter will make sure there are no hidden pathologies hard to detect and dependent on other curative methods.

6.5.1 Role of Barrier[13]

Clay is relatively watertight when it is applied in a continuous pasty layer. The quasi-parallel arrangement of the elementary particles one upon the other justifies the impermeability of clay paste to liquids and gases. Therefore air crosses this barrier only with difficulty and aerobic microorganisms can be fought directly by deprivation of gas, generally of oxygen, with a sufficient time of application.

This barrier effect could be assimilated to the particular beneficial effect observed on skin surfaces when covered with clay. Effectively superficial infections can be controlled by application of clay. It is necessary to note that people who work with clay, for example, potters, are rarely subjected to wound infection that they can contract (provided external causes do not bring germs to them). In addition to the

[13] In scientific papers one also finds the "barrier effect" which re-covers the same concept.

barrier effect, bacterial germs present on the skin can be adsorbed and trapped on the surface of the clay in contact with the skin which allows their elimination when clay is taken away.

More serious damage such as burns or ulcers[14] can also be fought thanks to application of clay taking good care of interposing gauze, fine paper, or sterile linen between the skin and clay. Doctors recommend that burns should be treated as soon as possible with cold water; the contribution of the clay pack can come only later. However, for second-degree burns or ulcers about to burst open, a medical approach is imperative: the doctor may recommend the use of clay, but it is especially inadvisable to resort to it a priori without medical opinion. Much medical evidence has proved the effectiveness of these treatments. In fact, clay does not cure wounds directly but provides a protection that reinforces natural defenses in two ways. The barrier effect is initially a physical mechanical effect: clay isolates from the external medium and maintains the humidity level necessary for tissues but without maceration. The second effect, less easy to describe, utilizes the reserve of exchangeable chemical elements from which the body can possibly benefit when it is locally in a very disturbed situation or chemically imbalanced. In this type of use, some perseverance is necessary until the wound is beginning to heal so that the surface part of the damaged tissue builds up a new barrier.

The same role can be visualized for the uses of clay by the digestive tracts. Two processes coexist and cannot be separated. A first action is to cover the walls of the digestive system and isolate them from the acid fluids that cause suffering. A second action is the constitution of a very complex gel with material contained in the stomach, then in the intestines. Clay then is in a very acid medium, which contains many polar molecules; this potential activity mobilizes all the mineral reaction capacities. Sometimes the sensitivity to acid leads to the destruction of clay by acid attack. The mobilization of the acid by clay gives a certain respite to the organism and relieves it. This action is rather often regarded as sufficient by patients who do not feel the need for medical intervention; the latter is, however, necessary when the symptoms persist and it is imperative in the event of important deterioration of the protective layer of mucus of the mucous membranes or in the case of lesion of the mucous membranes themselves.

In this digestive context, the clay mineral has a character of neutrality with respect to food because it is not a nutrient. However, in certain cases the body can perceive it as a sufficient food load to allay the feelings of hunger and thus avoid the absorption of food that will possibly be rejected. On this purely mechanical level, clay has allowed in the past, and still makes possible, preserving the morale of certain populations in periods of food shortage. Animals also resort to it in a spontaneous way when food is becoming scarce.

We have dealt here only with the barrier effect but it is also necessary for us to take into account the potential impact on the body of elements or molecules adsorbed on the layers of minerals and introduced in an indirect way at the same time as clay.

---

[14] By the use of phyllosilicates care concerning serious ulcerous states such as Buruli ulcers can be approached. The results are sufficiently convincing to justify more complete clinical trials.

6.5.2 Matter Brought in by Clay

The more common example is the reserve of water resulting from the capacity of clays to fix a more or less large quantity of water. The hydration of the layers can take the forms we described previously; we retain here only the fact of the presence of a quantity of easily exchangeable water, ready to be released from its substrate. When an organism suffers from a lack of water, the water contained in hydrated clay is likely to be sufficient to meet the immediate needs. The rehydration that results from it can save many people at little expense: only a small quantity of correctly hydrated clay that one can rather easily get in most situations (local clays are often sufficient).

Animals in the wild use clay spontaneously when they need water. Elephants, for example, can "find" water by trampling a piece of ground that seems apparently to be dry. The vibrations due to the repeated constraints support the mobility of water, then its insertion between the clay layers: it then mechanically presents a property of thixotropy that leads to the formation of viscous mud. Water can be thus brought back to a state of sufficiently high concentration to be usable through this mud which thus brings its water to the dehydrated body.

Plants also use this water contained in clay and the presence of clay in soils is essential to their good performance. In the water equilibrium, there is a spontaneous mechanism of balance between the forces of adsorption of water on the layers and the forces of capillarity due to the fine interparticle porosity of clays.

Water is the ideal example because it is universally present. However, the same mechanism of sorption to clay makes it possible to convey many other molecules at the contact of clay naturally or artificially. We cannot quote them in an exhaustive way. It is the mechanism that interests us here; in addition, it is the task of a physician and of a chemist to decide which molecule is to be used.

Note: The maintenance of a molecule in the interlayer space of clay is never to be taken for granted. There exists kinetics of saturation just as when this molecule penetrates into this space, there exists kinetics of desorption which is appropriate to it. Practically the study of these phenomena can be carried out in the laboratory by carrying out curves known as "isotherms of adsorption or desorption" that enable us to know the speeds of matter fixing or relaxation (gas, vapors, etc.) on or by clay under well-defined conditions.

6.5.3 Exchange of Matter Between Clay and the Body

The exchange of matter between clay and an organism is very interesting and useful. The process implies that a physicochemical mechanism can develop between clay and the human body. It is of great interest because a large proportion of molecules present in the digestive tracts or deposited on the skin or hair are reactive to clay. The contribution of ions or molecules by exchange is an important mode of action that allows essential reactions inside the very tissues and cells.

Minerals (clays included) and the human body are chemical systems having in common, in their constitution, the same chemical elements, although in different concentrations. Over time man has gained experience about the beneficial and hazardous effects of minerals *s.l.* (*senso latu*, i.e., natural inorganic and crystalline solids and natural chemical elements of inorganic origin) on health.

As a function of their concentration, the elements present in the human body, as also in minerals, are usually classified as major (such as oxygen, hydrogen, carbon, sulphur, phosphor, sodium, potassium, calcium, and magnesium), minor (such as iron and silicon), and trace elements (such as zinc, manganese, selenium, copper, strontium, fluorine, and iodine). In regard to human uptake of these chemical elements, food, water, and air are the sources of these chemical elements, and ingestion, inhalation, and dermal absorption are the intake pathways.

The contributions of trace elements are often evoked. It is obvious that those can be conveyed in exchangeable positions in clay in very varied forms. For these elements, it is enough that an exchange is actually carried out on some of the sites concerned to constitute a considerable and sufficient contribution to the organism.

Table 6.1 shows the average contents of the chemical element constituents of the human body.

6.5.4 *Extraction of Matter with Clay*

Fixation includes various processes (adsorption and absorption) that can justify the anchoring of ions or molecules on the mineral particle. The fixation of a molecule or a particle on clay can thus be seen; for example, greasy substances can be

Table 6.1 Chemical elements (major, minor, and trace) content (in g kg^{-1}) in the human body according to Schroeder

Major Elements of Human Body (g kg^{-1})		Oligo-Elements of Human Body (g kg^{-1})	
Oxygen	624.3	Zinc	0.033
Carbon	211.5	Rubidium	0.0046
Hydrogen	98.6	Strontium	0.0046
Nitrogen	31.0	Bromine	0.0029
Calcium	19.0	Copper	0.001
Phosphorus	9.5	Vanadium	0.0003
Potassium	2.3	Manganese	0.0002
Sulphur	1.6	Selenium	0.0002
Chlorine	0.8	Iodine	0.0002
Sodium	0.8	Molybdenum	0.0001
Magnesium	0.27	Nickel	0.0001
Iron	0.06	Chromium	0.00002
Silicon	0.02	Cobalt	0.00002
Fluor	0.037	Uranium	0.000001
		Beryllium	0.0000003

Favier A, in http://www.sante.ujf-grenoble.fr/sante/alfediam/nutrition/ol-6.pdf

difficult to eliminate from the skin, such as the dirty oil of mechanics or ink used by printers. Clay is extremely effective to fix these molecules and to remove them. Here we are in a field rather close to cosmetic care.

6.5.5 Transport of Matter Between Clay Particles

In this case, the matter is located between the clay particles (and not between the layers) and accompanies them in their movements. For liquids, it is obvious that they find a place in the interparticle clay porosity. Generally neutral and inactive solid matter particles are simply mixed mechanically with clay minerals, without detectable specific relations with them. It is a current case, and clay then has a driving role of the associated phases or to put it differently, it has a role of vector. It is necessary, however, that the quantity of clay should be sufficient so it can play this part.

Dermatological or capillary actions can be governed by this mechanism, it is the well-known example of body gumming.

6.5.6 Physical Action—Heat Transfer—Mechanics (Pressure)

We specified above that the heat storage capacity of much hydrated clay is an important physical property. It can occur in two directions: contribution or capture of heat. The contribution of heat is important to bring relief to a patient. Thermal extraction is useful in the event of high fever and can be associated in this case with a contribution of water and ions within the organism.

In addition, clay employed as cataplasm also has an excellent resistance to distortion and tearing making it possible to carry out soft but constant and flexible compressive actions that can act on blood circulation or traumatisms, for example. If this clay dries up and hardens in places, such as on an articulation that it wraps completely (for an anti-inflammatory effect), it acquires a sufficient rigidity to constitute a temporary immobilization with the advantage of being able to be broken up easily, which can prove to be invaluable in case of urgency.

References

Droy-Lefaix MT, Tateo F (2006) Clays and clay minerals as drug (Chapter 11.6). In: Handbook of Clay Science, Elsevier, Amsterdam, pp. 743–752

Duchaufour Ph (1997) L'abrégé de pedologie: sol, vegetation, environnement. Masson, Paris, Collection: Sciences de la Terre

Hénin S, Gras R, Monnier G (1969) Le profil cultural. Masson, Paris

Pomerol C, Ricour J (1992) Terroirs et thermalism de France, éd. BRGM

Van Damme H (1994) Physique et mécanique des boues et des glaises, rapport DGA n°90/851

Veniale F (ed) (1996) Atti di Convegno Argille Curative. Associazione Medica Italiana di Idroclimatologia, Talassologia e Terapia Física, Gruppo Italiano AIPEA. Salice Terme /PV, 145 pp

Veniale F, Barberis E, Carcangiu G, Morandi N, Setti M, Tamanini M, Tessier D, Barberis E, Carcangiu G, Morandi N, Setti M, Tamanini M, Tessier D (2004) Formulation of muds for pelotherapy: effect of "maturation" by different mineral water. Appl Clay Sci 25:135–138

Yvon J, Ferrand T (1996) Preparation ex-situ de peloids: Proprietés thermiques, mécaniques et d ´échanges. Proc. Meeting "Argille Curative", Salice Terme/PV, Gruppo Ital. AIPEA, 67–74

Chapter 7
Therapies Based on Clay

Abstract The therapies based on clay dealt with in this chapter are classified in two main groups, physical therapies and chemical therapies, involving either physical or chemical exchanges between the clay- bearing material and the human body, respectively. Physical therapies involve energy exchanges that can be induced during a massage by employing mechanical pressure on the previously heated (maximum at 45–50°C) clay pack, and from the pressure and heat being applied; the beneficial effects of improvement of muscular tone and blood circulation through abrasive and gumming action may occur. Chemical therapies involve chemical exchanges that could be promoted and developed by the applied pressure and heat. Beneficial effects could result from the chemical exchanges of the transdermal absorption of eventual bioessential elements and compounds existing in the liquid phase.

Historically the use of clays made it possible to develop empirical processes with real effectiveness, which justified the transfer of knowledge from generation to generation. Little by little these applications proved to be in relation to scientifically explained properties. Whenever possible we present these relations by specifying that the scientific projections can constantly justify an evolution of these models and consequently modify, make more accurate, or refute these explanations.

The two large fields of action we have just quoted above bring into play either the environmental relations (soft mechanical action, heat transfer) or the specific qualities of the clay particles (exchanges of water, chemical effects). Another type of action, more indirect but considerable, is the impact on the microorganisms. Clay can isolate them from the medium that contains them, coat them or destabilize them and neutralize their effects. It can also intervene by fixing toxins produced by these microorganisms and in this way suppress their effects.

7.1 Physical Therapies

In this category we put all the uses that bring into play a direct energy action (mechanical or heating) at the time of physical contact between the organism and an important mass of clay. Mud baths are listed in this chapter for part of their effects.

M. Rautureau et al., *Clays and Health*, DOI 10.1007/978-3-319-42884-0_7

7.1.1 Active Physiotherapy

7.1.1.1 Physical Energy Exchange; Mechanical Pressure

The physiotherapist needs to soften the manual contact when a prolonged massage causes an action of friction. For that he can use oil, water, or creams but also pastes containing clay.

In this type of application clay presents the very important property of deformation without rupture called plasticity: the particles slide over each other and the mass becomes deformed with very little friction. The result is comparable to a lubricating effect and gives the patient a soothing feeling. In addition, the individual deformations of the very many particles forming the paste add up. The paste can thus adapt itself to the shape of the surface supporting it and so ensure a perfect marriage of the space between the surface of the hand and the treated area of the organism. Plasticity depends on the content of the liquid and its nature.

The variations of local energy balance can cause ruptures in the continuity of the clay mass. The forms of these ruptures are characteristic and carry information on the state of dehydration of the paste, with the image of the withdrawal cracks when a clay soil becomes too dry.

The fact that particles built of some layers keep their coherence causes us to take into account the concept of internal energy of this system. After application, at the interface between clay and the organism, the energy of the clay mass must find its balance, such as the surface tension energy that develops between three components (e.g., liquid, solid, and gas). Thus there is not only one simple correspondence of form between the substrate (clay paste) and its support (skin) but also an energy equilibrium that confers an adherence between these two media, favorable to the transfers.

Qualitative but also quantitative methods are beginning to be described in this context. Inasmuch as modeling studies are not easy because many parameters come into play, researchers then chose a traditional preliminary method known as the "indices method." It consists of combining several active parameters whose important action is known. Two of them are now in use giving expression to the quality and applicability of thermal mud or peloid (see *Handbook of Clay Science*, Chapter 11; Bettero et al. 1999):

1. Index RTM®[1] assesses the evolution of viscosity, plasticity, and of textural and structural features of the thermal mud/peloid paste.
2. Index TSV®[2] assesses the bioadherence or tensiometric affinity of the thermal mud/peloid paste to the skin in the contact zone; biometric data is used in cosmetics to understand the affinity between skin and the cosmetic products.

---

[1] RTM®: Rotary Transfer Mechanism: Index that represents the evolution of the viscous, elastic, viscoelastic, and structural features of the mud/peloid.

[2] TSV®: Tensiometric Versus Skin: Bioadhesive index that represents the affinity of different mud/peloid for the surface of a certain skin.

However, these approaches are complex and lead to preliminary qualitative indications, often essential, but that must give way little by little, in the near future, to quantified studies and more robust models in the framework of the theory.

7.1.1.2 Exchange Water

The clay paste is very often made with water. The quantity of water can vary up to a point without the property of plasticity being basically disturbed or lost. The paste can thus be used with distinct degrees of viscosity.

The homogeneous water–clay mixture is interesting from another point of view because it makes it possible to some extent to have a sort of supply of extremely useful liquid to avoid skin dehydration. For example, in certain cases of organisms in a state of advanced dehydration, this available and exchangeable water allows a gentle rehydration thanks to the functioning of the type of "osmosis membrane" specific to the skin.

The water of the water–clay mixture is characterized by its connections with the clay particles and by its wealth of chemical elements whose presence is natural, but can also be modified voluntarily during the preparation. This mixture, although complex, is a reserve on which the organism can draw according to its needs during the application. All the know-how of the user is needed to manage the concentrations of the involved elements, which is sometimes rather difficult.

The chemical elements of the hydrated clay paste can have important effects that must be permanently kept in mind for many uses.

During the application of the paste it is necessary to supervise the characteristics of the mixture that can evolve:

1. The cooling rate of the paste
2. The atmospheric evaporation
3. The local reduction of the hydration state of the clay–water mass due to the absorption of water by the body

Compared to the availability of water, *a contrario*, dry clay, thanks to the property of adsorption and according to the conditions at the time of employment, can constitute an active desiccant, good or harmful according to the case (e.g., talc).

The evolution of the paste's rate of hydration may alter its plasticity and its coherence. It should be prevented from becoming hard or breakable by controlling the hydration state to preserve its optimal effectiveness. In practice this is know-how that is acquired by repetition and advice. The empirical aspect of this type of intervention is never completely controlled and must not escape us.

7.1.1.3 Heat Transfer

Pastes and muds contain a great quantity of water whose heat storage capacity[3] is high (equal to 4.186 J K^{-1} kg^{-1}). Consequently a clay paste can contain a significant amount of heat. The strong heat capacity and the bad thermal conductibility of the paste, due to the structure in layers of clay that imprison water, allow a long conservation of this heat. On the other hand, the increase in initial temperature is rather slow; it is thus appropriate to control it well.

Neither the body nor the clay mineral must be disturbed by temperature. The presence of water makes it possible to limit and control the temperature at bearable levels for the body. In the same way water makes it possible to ensure that the clay is not deteriorated and does not lose its properties. The transfer of heat to the body could also be carried out by blowing hot air on it. However, to pulsate hot air dries the target and in our case the body of the patient. This water loss of the body by evapotranspiration can be harmful. The body's dehydration is avoided by using a warm clay paste that enables contact with a significant amount of exchangeable water while ensuring a reasonable thermal contribution. It is a subtle nuance, but the difference of the actions is fundamental and gives a solid justification for the use of clay. Moreover, the rehydration is progressive; the method is applied in dry and hot countries when a body (human or animal) has undergone a loss of water dangerous for its survival.

In the case of a body's contact with clay—a poultice, for example—we must remember that the control of temperature is always necessary in order to avoid burns. The water contained in a paste is, indeed, likely to maintain too high a temperature for a sufficiently long time to cause these burns.[4] The temperature must be associated with the concept of "exchangeable thermal mass" which leads to taking into account, during a heat exchange, the respective masses of the donor and receiver and their respective heat storage capacities. Balance is obtained on the temperature. It should then be ensured that the quantity of transferred heat does not cause the patient too great a rise in temperature, which can be dangerous.

In connection with the behavior of the clay mass depending on temperature, we must not forget that the latter influences the strength of the bonds between water or various other components and clay. The transfer properties also vary. It is a

[3] The heat storage capacity mass C_P of a material represents the quantity of heat ΔQ necessary to raise by 1° the temperature of a unit of mass of this material. It is expressed in joules per degree and per kilogram (J K^{-1} kg^{-1}) where the symbol K represents the Kelvin degree and kg the kilogram.

A convenient unit that has been employed for a long time is the "calorie" which is the quantity of heat making it possible to raise a mass of 1 g of pure water from 0 to 1°C. This unit should not be used any longer; however, it is related to a quantity of fundamental heat for all living organisms containing water.

[4] Let us recall that doctors consider that a 3 s contact of the skin with water at 60°C is enough to cause a (very) severe burn to a young child who does not have fast enough reflexes to get away from it. Current legislation proposed a safety limit, however, still high, which is at 50°C for domestic current hot water.

complicated field to analyze but which must be taken into account especially in case of failure. For example, it is well known that the effect of temperature on the effectiveness of the clay applications can, depending on the case, lead to stimulating or alleviating results. A medical rule among many others is not to introduce heat in case of blood discharge which might be confused with the painful effect of a contusion.

7.1.1.4 Regulation of the Muscular Tonus and Circulation

We are interested here in the various concepts of gravity, pressure, and tension. These three concepts are common to us. We cannot avoid gravity in our normal living conditions on Earth. Gravity causes anomalies that can generate true pathologies. Muscular tonus and blood circulation are directly concerned. Simple atmospheric pressure is not enough to compensate the effect of internal pressure due to verticality of the body. Immersion in a mud bath of clay creates a compression that, without removing the phenomenon of gravity, makes up for it partly or entirely by the phenomenon of pressure related to Archimedes' principle. The result is a physical relief that supplements the effects of the other properties of clay.

During the coating of a part of the patient's body, mud creates a compressive force larger than that of water. This useful property in the treatment of blood circulation or muscular tonus is put into practice by some balneotherapy establishments.

We saw that the clay paste is plastic. The curvatures and deformations imposed on clay during the applications create internal forces with the pasty mass. These forces appear at the interface with the body by a mechanical tension comparable to the surface tension for a liquid. To break the continuity of the paste, it is necessary to impose a relatively important force. Nevertheless this paste is manually deformable thanks to acceptable efforts by the body that undergoes them, and which in general perceives them as relatively soft and continuous. It seems possible to use this property when applying a paste to rebuild wounded tissue (e.g., cicatrization).

The behavior of a paste subjected to efforts is extremely complex. Researchers endeavored to understand its fundamental aspects. Thus, knowledge of the mechanisms of resistance to the deformation and the mechanism of penetration of a fluid in a paste could be invaluable for the users and medical practitioners using clay (Van Damme 1991, 1994).

7.1.1.5 Abrasive Effect, Adherence, and Gumming

Although clay is considered soft, it should not be forgotten that on a microscopic scale the particles can be mechanically resistant and physically hard. Microcrystals of some varieties have prismatic forms (in particular, kaolinite and illite) whose contours are clear-cut and quite visible in scan and transmission electron microscopy (see Fig. 7.1). In contact with the skin, acting on soft tissues, these microcrystals can have a sharp effect at this scale. However, the extreme thinness of the clay

Fig. 7.1 Images of particle aggregates and of free particles; kaolinite particle aggregates observed with SEM (**a**); diagram of the particles (**b**); illite, flexible fibers observed with SEM (**c**, **e**) and with TEM (**g**); kaolinite, rigid particles observed with SEM (**d**, **f**) and with MET (**h**). TEM transmission electron microscopy, SEM scan electronic microscopy

particles is not very dangerous and on a macroscopic scale cannot cut or wound. Let us note that fibrous clays must be absolutely excluded from medical use.

From another point of view, clay particles have a specific large surface and carry electric charges (to compensate for those due to heterovalent substitutions); this situation causes phenomena of adherence.

By interaction of these two mechanisms we have a double possibility of action: *abrasion*[5] and *gumming*.[6]

For capillary care, for example, these two techniques can be implemented. Hair has diameters ranging from 45 to 100 μm; its surface is made of keratin scales. This dimension offers a completely suitable surface so clay particles can effectively exert an abrasive effect during friction and physically rid the surface of the hair of the extraneous particles that are stuck there as well as the damaged or ageing parts of the hair. In addition, the abrasive action of clay is often combined with its fixing and exchange properties of coloring, rebuilding, or protection molecules.

We find, in body aesthetics and care of the skin, many applications of these two methods. In almost all cases, the action of clay utilizes the transport and exchange of molecules or particles. Their use renders many services, always appreciated and of a great reliability.

Figure 7.1 shows the high porosity developed when clay particles are gathered in aggregates (a). The free spaces existing between the aggregates and between the constituent particles make sliding movements possible between all of them, allowing the possibility of the whole assemblage to become deformed, and consequently the plasticity of clays.

7.1.1.6 Plastic Deformation and Sensory Perception

The clay pastes have a resistance to deformation (plasticity) that varies according to their water content. Paste interposed between the hand of the masseur and the skin of the patient constitutes a sort of buffer that moderates the constraint when the plasticized paste moves under the imposed effort. The paste becomes deformed and is adjusted with that of the zone of the body which it covers. The force is practically constant as long as a significant thickness of paste is available and active. This constancy results from the property of plasticity. Plasticity remains a stable property of the clay paste as long as its hydration rate remains between that of a dry material (it is then almost solid and breakable) and that of a fluid material (it loses its shape and flows under the effect of its own weight).

A deformable material is plastic when it keeps its coherence when a force is applied to it. The material becomes deformed according to available space and obstacles it meets without undergoing fractures. Complex studies were developed

[5] Abrasion is caused by the extraction of matter by wear associated with a sharpening effect resulting from a difference in hardness and an adapted geometry.

[6] Gumming is the removal on a mobile support of a fraction of matter fixed on a surface by adherence.

on the type, amplitude and mode of constraints necessary to the fracturing of a pasty clay mass (Van Damme 1991). This theoretical knowledge is essential to develop new applications.

The plasticity property can be useful to carry out physical exercises requiring personal and self-controlled efforts and for soft muscular care rehabilitations. In other cases perception rehabilitation by contact and touch is possible by using clay, for example, after reconstitution of very fragile tissues following burns, serious dermatological aggressions, or surgical operations.

In other fields such as autism, the very particular physical contact of clay with the skin can be an appreciable help for the training or re-education of perception. The evolution of the form of a mass of clay paste by the pressure exerted by the patient herself implies a voluntary act.

Clay paste is also used in art therapy. This material then presents a double possibility: on the psychotherapeutic level and on the actual physical care that can be associated there (forces, muscular restructuring, hydration, contributions of elements, etc.).

If the quantity of water present in a paste gradually increases, the property of plasticity is lost and the property of fluid viscosity benefits. This viscosity can vary in a sufficiently wide range to lend itself to mechanical applications, for example, in the field of the physiological rebuilding of strains or quite simply for maintaining muscular condition.

7.2 Peloids, Cataplasms, and Bandages

We kept the word *peloid*, barely known to the public at large, but usually used in the medium of hydrotherapy. By definition it is "a mixture of mineral water or sea water[7] associated with mineral or organic matters." Thus this term covers a vast domain, but for the applications that interest us, we keep in mind that a peloid can be formed starting from clay. In addition, it is what happens when kaolinite and/or smectite is used. These clays are generally extracted from deposits close to the place of use (autochthonous clays). Many fundamental studies have been done on this technique, among them the articles of synthesis by Yvon and Ferrand (1996), Veniale et al. (2005), and Veniale et al. (2007), which also give extensive bibliographies.

The main technical difficulty in making a clay paste is to obtain a good homogeneity of the mixture because clay at the beginning of its humidification becomes almost impermeable to the spontaneous diffusion of water, at least in a reasonable time. When a great quantity of clay is to be prepared without mechanical intervention, the impregnation will require a very long time (several months[8])

[7] Seawater, physiological salt solution, and transfer of ions per dialysis often intervene in the mechanisms of action between clays and organism.

[8] It is well known that dryness causes cracks on the surface of clay soils. Farmers know that under these conditions, the return of the ground in a satisfactory hydrometrical state can be very long, sometimes several years.

when the preparation of more than a few kilograms of paste is required; this is not very compatible with domestic and family preparation. As regards the quantity of water, too much of it should not be introduced, under penalty of losing its paste character and therefore its plasticity. In this case, to return to a suitable plasticity is very slow and relatively difficult to control. It is a delicate operation that requires a proven practice to get an effective product.

The limit between paste deformable only by one sustained effort, and mud that presents the property of flowing under its own weight, does not depend only on the fluid quantity present but also on its nature. The pH of the medium is an important datum in this distinction. The behavior of clay introduced to stabilize a digestion problem brings into play its aptitude to form a medium having the required characteristics of fluidity. It is the aptitude to form a gel that interests us here. The use of clay makes it possible to take, in part, control of the pH of the extremely complex unit made up of the mineral mixed with the stomachic or intestinal contents.

According to Yvon and Ferrand (1996, op cit.), a good peloid has three qualities: slow cooling, facility of handling, and pleasant contact with the skin. A viscosity of 400 poises is generally retained so that the paste is easy to handle while keeping a satisfactory plasticity. After use, the peloid applied must offer the possibility of being withdrawn easily without sticking to its support.

To obtain a viscosity of 400 poises,[9] the water content of the paste varies strongly according to the nature of the clay, of 95% for a kaolinite with 400% for a bentonite. The correct identification of the minerals as well as thorough knowledge of the initial product is thus necessary to prepare in a homogeneous way important quantities of paste presenting the required properties.

Mixing pastes leads to their equilibrium and homogeneity. Several months are theoretically necessary to reach this state, but producers shorten this unacceptable duration to approximately 20 h, sometimes followed by a somewhat long period known as "maturation." The state obtained is controllable thanks to the property of swelling measured with an oedometer (Yvon et al. 1993). At the end of the maturation time, the obtained paste places the clay in a physicochemical state that can be regarded as a state of "dynamic balance"[10] which makes it "active."

(Cataplasms use a clay paste, therefore a peloid even if there are only very few added or modified components). They are applied to the skin at a temperature adapted to the considered pathology. For congestion, torn muscles, or inflammation, a cataplasm must be cold. To tone up the body it is used tepid, without exceeding an acceptable temperature by the organism under penalty of burns as we have written above (in the vicinity of 40°C for the skin of a young child). Parallel to this concept of direct perception of heat, clay and the organism exchange elements in ionic or molecular form; this energy exchange gives the patient a feeling of warmth and well-being.

---

[9] Poise: Traditional measuring unit characterizing dynamic viscosity. This name comes from the French physiologist Dr. Jean Poiseuille (1799–1869). Currently this unit is replaced by the legal unit: Pascal second (Pa s) with the correspondence 1 Pa s = 10 poises.

[10] We regard as "dynamics" a balance likely to evolve/move when the external medium is modified.

The property of exchange, ionic or molecular, with the treated organism requires putting clay in direct contact with the skin, without intermediate fabric, or if absolutely necessary, by using a piece of sufficiently permeable gauze.

The duration of application of the cataplasm is an important parameter that allows the effectiveness of the various transfers (chemical element and heat transfers). The draining of the thermal reserve as well as the evolution towards the balance of the chemical concentrations must be permanently taken into account.

7.2.1 Aesthetics and Well-Being

Skin is sensitive and fragile; it can be very easily damaged or soiled by external elements. It is initially a frontier between two media but also the place of transmission of many perceptions and a place of exchange of varied matters by diffusion.

As a rule body care is generally accompanied by any action giving the body the most complete relaxation possible. The softness of the contact with the majority of clays completely satisfies these objectives. The water contained in the clay paste has an important heat storage capacity; it slowly restores the heat it contains at the time of contact with the body. In addition the presence of water guarantees maintenance of the state of hydration of tissues located near the treated zone. These dynamic hydrothermal methods are very interesting and largely widespread in thalassotherapy.

When clay pastes are being used it is necessary to take into account the impact of the presence of particles of very small dimensions (clay or associated minerals), sometimes nanometric, that have a capacity of very strong adherence. If they remain on the skin after removal of the application (peloid or masks) they can have a drying effect. It is one use, inter alia, that justifies the use of powdery talc to dry the skin.

7.2.2 Softness Effect

Once more let us recall that plasticity specific to the clay pastes is due to the layer texture of the particles. The presence of water in the clay paste decreases the forces coming from the connections and interparticle frictions. Under low constraints, the particles can slide one over the other, making it possible for the paste to become easily deformed without rupture and also to follow the smallest muscular movements that are inevitable even in periods of voluntary relaxation. It gives an impression of comfort through the absence of constraint, continuity, and softness of the contact.

For dry use, much rarer, we already quoted talc. This mineral has almost no atomic defects of substitutions; its layers are thus neutral and consequently they naturally slide one over the other.

7.3 Therapies Involving Chemical Exchanges

It is also necessary to consider the classification of the phyllosilicates in order to identify correctly the nature of the chemical elements present in each clay mineral, species, or variety. For more precise details, one may consult Caillère et al. (1982), Rautureau et al. (2004), and Guggenheim et al. (2006).

Thus it is possible to note the universal presence of silicon in the tetrahedral layers as well as aluminum, magnesium, or iron in the octahedral layers. Isomorphic atomic substitutions justify the presence of very interesting elements such as cobalt, chromium, or lithium and others, which some users can compare with the list of the trace elements that can be present in clay. However, we point out the possible risks of too hasty a generalization of this matter (which results in confusing the structural elements of a clay mineral with those from other mineral phases associated with it). These elements are really invaluable and beneficial for human biological stability. In addition, oxygen and OH^- hydroxyl groups present in all clay mineral structures are frequently replaced by small amounts of fluorine and it is well known that this element is also very important for health, even if it is rarely quantified. Let us note that all these chemical elements are incorporated in the crystal lattice of clay minerals, and if we want to benefit from them their extraction requires disruption of the clay mineral structure, which is not easy.

7.3.1 Generalities

A clay particle is, as a rule, a flat microcrystal resulting from the superposition of parallel elementary structural layers that are themselves constituted of structural sheets (Fig. 6.1). Also, it is necessary to consider both the external crystal surface consisting of the so-called basal, the edge surfaces, and the internal surface consisting of the so-called interlayer surfaces, as well as the layer mass itself, which all contribute to clay particle specific properties.

Clay particles and their elementary layers are sensitive to the conditions of the external medium and particularly to its pH (which defines the acid or basic character[11]). It is necessary to be quite aware that the octahedral and tetrahedral layers are not equivalent (see Chapter 2). Thus, we know that an acid attack manages partially to extract the octahedral cations under conditions which are not very different from those of the stomachic medium. The octahedral layer is the more fragile, and can be gradually destroyed by acid attacks. The release of ions from this layer, often aluminum or magnesium, but also iron, can have an important incidence on the human body (e.g., aluminum is known as a neurotoxic element associated with disordered

[11] The "hydrogen potential" or pH varies between 0 and 14, a neutral medium has pH = 7, an acid medium has pH lower than 7, and a basic medium has pH higher than 7. The value of the pH is related to the concentration in H^+ ions, noted $[H^+]$, by the formula: $pH = -\log_{10} [H^+]$.

states close to encephalopathy). The siliceous residue of the tetrahedral layers remains after acid attacks of the clay mineral. This "silica" can show much varied forms inherited from initial mineral characteristics, and this fact could be important. The structure of this residual silica, for example, in the case of sepiolite, is partially fibrous (Gouami 1996).

Clay's action on health is often attributed to a contribution, either through food or through contact with the skin, "of the chemical elements known as natural" beneficial to health and for which the psychological approach is not to be neglected. Traditional trace elements (see Table 6.1) are well known, but what do they represent in fact, and how useful is their contribution? It is necessary to go very far in the analytical field to explore these very particular properties of clays. In fact, the majority of the mentioned trace elements come from materials associated with clay. However, some of them are also likely to come from clay minerals themselves, after direct exchange or the effect of an acid attack, for example.

All the chemical elements shown in the periodic table, with the exception of rare gases and artificial chemical elements, can be found in clays. We cannot treat in an exhaustive way all the interactions that bring the chemical elements into play. Nevertheless we detail some of these points below.

7.3.2 Nature and Position of the Exchanged Elements

Water is omnipresent in and around clay. Its various forms (free hygroscopic water and interstitial water, zeolitic water, structural water,[12] etc.) have distinct binding energy to the layers and, consequently, distinct mobility. When clay is included in a reactive medium, exchanges involving these various types of water are possible. This property is important because clay then becomes either a hydrating or a desiccating agent that modifies the water concentration of the medium to which clay is applied. Note that the change of the hydration state of the mineral paste is reflected in its physical properties, for example, in its plasticity or capacity to remain in suspension.

In a dry state, clay particles gather in aggregates. Between clay particles the existing pores are ready to receive, among other possibilities, a liquid that can create connections with mineral surfaces. The forces of these connections are in general rather weak. However, when the quantity of liquid is important, these forces can be sufficient to form a stable gel with very little mineral matter. (This property is called thixotropy.)

Another category of space, also available for the exchanges, is located inside the particle itself, between the elementary layers: it is also named interlayer

[12] "Zeolitic" water is characterized by weak bonds in the mineral structure, generally trapped in channels or cavities specific to certain minerals. Under normal pressure conditions this water is eliminated between 100 and 105°C, a fact which proves that this type of water is not completely free, but weakly linked to the mineral structure or by solvation of the interlayer cations.

porosity. At this level, forces resulting from the electric charges due to the chemical composition defects are much more intense. To compensate for this electric imbalance between the layers, ions or molecules become strongly bonded to the layers. These ions themselves can be surrounded by charged particles and/or polar molecules, water, for example. In that case we talk of hydrated compensation ions that decrease the attraction force between the layers and create new properties. The situations of the unit's imbalance, in a liquid medium, can result in a relaxation of the bonding strengths of the hydrated ions related to the layers that thus move away from one another. The result is a possibility of release of these ions in the external medium. These imbalances can occur at the time of setting the paste or mud in contact with the skin or by stomachic ingestion where the pH is very variable.

Electrostatic forces created outside the layers by the nonisomorphic substitutions located within the layers vary according to the tetrahedral or octahedral structural position (or from both at the same time) of these ions. In the interlayer position the most current compensation ions are: Na^+, K^+, Ca^{2+}, H^+, or OH^- and sometimes Fe^{2+}, Fe^{3+}, Al^{3+}, Mg^{2+}, Co^{2+}, or Cu^{2+}, among others. However, many other ions can intervene, either in natural situations, or after the intervention of physical chemists or chemists. But, right now, we can become aware of the immense variety of the possibilities and among them the important source of ions that are thus available for the body. Some of these ions which are mobile because of their structural situation belong to the group of the trace elements and thus join the field of properties of the latter and of health.

For the patient, an analysis of his or her specific personal needs must be made before any treatment by a physician.

If clay introduces ions, the opposite mechanism is also possible. The internal defects in the layers are compensated by exchangeable ions. This exchange mechanism is also a means of moving and extracting the ions coming from the body; it should never be neglected.

7.3.3 Kinetics of the Exchanges of Molecules by Clay

The origin of the ionic exchange capacity of clay (see Table 5.9) is due to the atomic defects of substitutions that it contains. During the ionic exchange the available quantity of ions of a given type varies. The kinetics of the exchange differs according to whether there is a very dense medium (clay paste) or a highly hydrated one (mud or liquid suspension). In this latter case the speed of diffusion of the ions is high, whereas a paste opposes more resistance to their progression. It is usual that on the level of the contact zone with the body a paste becomes poor in active ions before the latter become renewed by diffusion inside the paste. A new mixture of the paste by kneading then proves necessary to homogenize the unit.

According to the law of mass action the tendency towards a balance is carried out whereas only a part of the ions is absorbed by the body. Gradually, the diffusion mechanism becomes slower. Clay loses part of its effectiveness and must be renewed. After use it is not reasonable to consider repairing the composition of clay by carrying out an opposite exchange; that would be too expensive and technically difficult. Nevertheless, this type of operation is perfectly possible and feasible for work in laboratory studies or for the initial preparation of controlled compositions of clays.

7.3.4 Limits of the Exchange Capacity of Clay

We have seen that clays have a great aptitude to attract and retain anions or cations. The first source of ions comes from the medium directly to their contact (in our case a body, via its skin and of a liquid medium). If the medium containing the clay has a sufficient reserve, the capacity of exchange of clay is gradually saturated by the same type of ion. A balance is established; it depends on the nature of the exchanged ions. For example, if we consider a calcium–sodium exchange, by adopting a simplified writing,[13] we can then have schematically for ionic balance in an aqueous medium:

$$\text{clay-Ca} + 2\,\text{Na} \rightleftharpoons \text{clay}\Big<\begin{matrix}\text{Na}\\ \text{Na}\end{matrix} + \text{Ca}$$

The exchange of ions is established in order to lower the stronger ionic concentration until it reaches a balance if the ion reserve and the duration allow it. If saturation is obtained all the charged sites of the clay are then compensated by the same type of ion; the clay is known as monocationic. In this case, the measurement of the quantity of ions characterizes the theoretical maximum exchange capacity. Practically the clay particles consist of several layers that, according to the state of hydration and dispersion, do not all take part in the exchange reaction and do not always allow the complete realization of the expected exchange. In this case because of the textural state of the particles the theoretical limiting value is not reached. The exchange can be improved by increasing the quantity of water and by employing mud or a very liquid suspension for which the circulation of the ions (their diffusion) is easier. These various methods are employed implicitly in balneotherapy.

[13] As mentioned above, the notations of the type Clay–Ca or Clay–Na mean that the layer charges are compensated by calcium or sodium ions, called calcic clay and sodic clay, respectively. This compensation is done with respect to the balance of the electric charges.

7.3.5 Risks of the Reciprocal Exchanges Between Clay and the Body

The exchanges between clay and the body are here wanted and justify the choice of a given application. However, there exist considerable reciprocal effects that should not be underestimated.

By carefully choosing the clay mixture composition we can manage to ensure needed relief to the body. In this search for balance, we should not, however, forget reciprocal effect and the possibility of impoverishment of one of the body's components to the profit of the clayey mass at the contact. It is a possible source of imbalance for the body.

7.3.6 Contributions of Trace Elements from Clay

This point deserves to be approached separately because it is often mentioned and appears on many commercial composition sheets.

We must consider that from the list of the trace elements named in the balance of the living organisms[14] some are present in the very structure of clays. In particular the most important (potassium, sodium, calcium, magnesium, and phosphorus) which are present simultaneously in the organism and the clays, can migrate from one to the other according to the respective concentrations.

Thanks to the deterioration of clay mineral layers controlled by the external pH, the chemical elements can be released from the crystalline structure in ionic form: Mg^{2+}, Fe^{2+} or Fe^{3+}, Al^{3+}, Li^{+}, F^{-}, and so on. These ions cannot remain in a free state; they recombine quickly.

Other elements do not come from the structure of clay but are present in compounds associated with it, because it should not be forgotten that very often clay is a material of detrital origin, finely divided, and seldom pure. Clay–water suspension, after decantation, makes it possible to find in the floating liquid (commonly called clay milk) elements that are dissolved or present in the form of very fine particles. From this point of view, it can be admitted that a contribution of trace elements can come from the use or consumption of clay (see the following chapter).

In this field, it is necessary to be careful and to refer to precise analyses that are generally not clearly formulated by suppliers. These analyses are very difficult to achieve because they are minor components. Undoubtedly some of these elements are present there; natural clay by itself cannot be regarded as having reliable and controlled source trace elements.

[14] The trace elements are: magnesium, fluorine, iron, zinc, silicon, copper, tin, iodine, manganese, selenium, vanadium, nickel, chromium, molybdenum, cobalt, arsenic, bismuth, chlorine, mercury, lithium, phosphorus, potassium, sodium, sulphur, uranium, aluminum, silver, and gold (see Table 6.1).

When the contribution of these chemical minor elements in clay is specifically required, it is necessary to obtain a medical opinion because they are associated with an important mineral mass of clay that can never be neglected.

From another point of view, some elements considered as trace elements for the body can be major elements in clay. It is the case of silicon which is strongly dependent in an oxygen tetrahedron that we previously described such as the SiO_4 tetrahedron and which is particularly stable. This silica shape does not seem to be assimilable as such by the body; it would be rather harmful. If these silica particles are too bulky they can cause silicosis. Nevertheless the body needs silicon, and it finds it in various forms, inter alia under organic forms. It is not a chemical element directly supplied by clay. This "organic silicon" is the subject of specific research.

7.4 Conditions for Use

There are two ways of tackling this subject. First, the parts of the body that are being affected (skin and scalp, digestive bodies, etc.) have to be considered; then it is possible to consider the type of pathology clinically suggested (infection, pain, dysfunction, etc.).

Obviously it is the patient who expresses himself by indicating the parts of the body concerned and then the nature of the disorder to be corrected if possible. A very clear distinction must be made according to whether the zone of the symptom is accessible by direct contact or if it is in an internal part that cannot be put in contact with clay. As we have already specified, one should not confuse the therapeutic direct action of clay on a body and those preparatory but complementary to the drug effect or with the recovery of the normal functions of the body; in that case clay has only a subsidiary role.

Several clay therapeutic uses are referred to in papers as alternative medicine methods. In the majority of cases, there is no follow-up of serious clinical studies. However, these applications have some economic interest, because of strong public demand. It is the case with water cure applications or developments in gastroenterology or dermatology (clinical or beauty care). In these fields real and solid scientific studies were developed. There is thus a great disparity of knowledge to which one can refer.

Actual practices, referring to ancestral practices, often use clay mixed with extracts of plants or products such as oils. It is then relatively complicated to distinguish the cause (here: clay, oil, or the mixture) of the effect. Properties of adsorption and exchange of clay can play a major role in the case of these associations. This leads us to a system with three parameters: body, clay, and added products.

References

Bettero A, Marcazzan M, Semenzato A (1999) Aspetti reologici e tensiometrici di matrici fangose di impiego termale e cosmético. Proposta di un protocollo per la loro qualificazione. Mineral Petrogr Acta XLII:277–286

Caillère S, Hénin S, Rautureau M (1982) Minéralogie des Argiles, vol 2, 2nd edn. Masson, Paris

Gouami A (1996) Des sépiolites sonifiées aux gels de silice fibreuse: propriétés physicochimiques, texturales et rhéologiques. Thèse, Université d' Orléans

Guggenheim S, Adams JM, Bain DC, Bergaya F, Brigatti MF, Drits VA, Formoso MLL, Galán E, Kogure T, Stanjek H (2006) Summary of recommendations of nomenclature committees relevant to clay mineralogy: report of the Association Internationale pour l'Étude des Argiles (AIPEA) Nomenclature Committee for 2006

Rautureau M, Caillère S, Hénin H (2004) Les Argiles, 2e edition, Société de l'industrie minérale ed. (Sim, 17 rue Saint- Séverin, 75005 Paris). ISBN 2-904845-30-5

Van Damme H, Lemaire E, Levitz P (1991) La physique de la boue, Pour la Science, 167:82–90

Van Damme H (1994) Physique et mécanique des boues et des glaises, rapport DGA n°90 / 851

Veniale F, Setti M, Tessier D, Carcangiu G (2005) Physical parameters of thermal muds ("peloids"): influence of "maturation" procedures. Réunion Groupe Français des Argiles, Paris, p 35, abstract

Veniale F, Bettero A, Jobstraibizer P, Setti M (2007) Thermal muds: perspectives of innovations. Appl Clay Sci 36:141–147

Yvon J, Ferrand T (1996) Preparation ex-situ de peloids: Proprietés thermiques, mécaniques et d'échanges. Proc. meeting "Argille Curative", Salice Terme/PV, Gruppo Ital. AIPEA, pp 67–74

Yvon J, Thomas F, Villiéras F, Michot LJ (1993) Surface activity of water. In: Julia D et al. (ed) Power technology of pharmaceutical process. Handbook of power technology, 9, vol IV. Elsevier Sciences Publishers BV, Amsterdam, pp 79–113

Chapter 8
Principal Modes of Clay Use

Abstract There is a long tradition of the use of clay for therapeutic purposes. The so-called healing or curative clay can be used either for internal or external applications. The so-called edible clay is a type of healing clay used for internal applications, ingested in the form of clay/potable water dispersions (argillic water), bits/pieces of natural clay, or even manufactured cookies or wafers consisting of a clay–animal fat blend and slowly chewed.

For external or topical therapeutic applications clay is used in the form of a clay–water paste called either mud (equivalent to natural peloid), or just peloid or peloid *s.s.* (*strictu senso*) that corresponds to maturated and manipulated mud or natural peloid, called mud therapy and peloid therapy, respectively. Mud therapy and peloid therapy have been successfully used as treatments for various inflammatory disorders including rheumatic diseases and chronic inflammatory skin disorders such as psoriasis or atopic dermatitis. The mechanisms of action of mud and peloid packs for healing and skin care have been and still are the object of study by physicians specialized in medical hydrology, the mechanisms being attributed to a combination of thermal and physicochemical effects.

We have approached above the large domains of clay application (skin and intestinal mucous) and the excluded domains of clay application (respiratory tracts, visual and auditory systems, internal actions within the organs). Also, the clay mineral properties that come into play in varied types of application have been mentioned. Now, without entering into the specifically medical field, which is too complex and outside our descriptive framework, we deal with applications that we wish to preserve.

8.1 External or Topical Application

8.1.1 Clay in Contact with Skin and Pelotherapy

Bringing clay into contact with the skin is easy. In theory, the contact with skin is direct. There is neither mucus nor diluting agent as in the case of the alimentary bolus. However, attention must be paid to eventual surface secretions (sebum and sweat). It should also not be forgotten that dead cells and bacteria are contained and retained by the surface of the skin.

M. Rautureau et al., *Clays and Health*, DOI 10.1007/978-3-319-42884-0_8

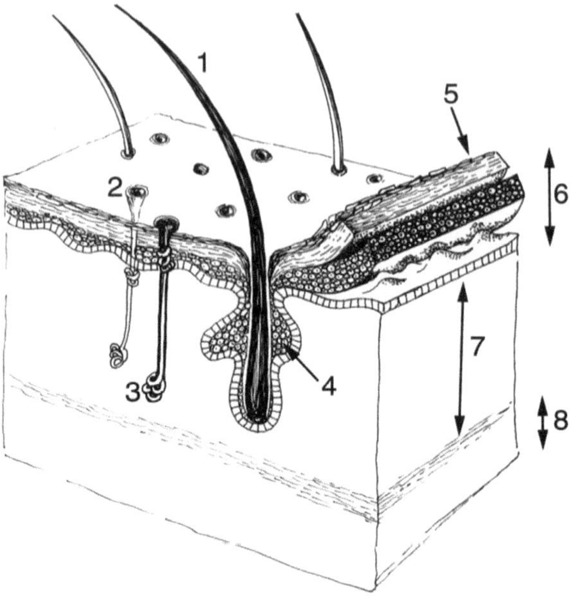

Fig. 8.1 Diagram of skin structure. *1* Hair, *2* Perspiration pore, *3* Sudoriferous gland, *4* Sebaceous gland, *5* Exfoliation zone, *6* Skin, *7* Derm, *8* Hypoderm

Skin contains internal glands that open to the skin surface by pores. The sweat glands secrete sweat to eliminate toxins and cause skin cooling by allowing water evaporation. Sebaceous glands secrete sebum which lubricates skin hairs and prevents the drying of the cutaneous surface. Clay reacts to these various products when put in contact with them. The surface of the skin is in fact the first zone of action constituting an interface. But there is a possibility of deeper action because the various layers of the skin are permeable to the elements brought in by clay. The most current of these elements is water, weakly related to clay particles by hydrogen bonds. Then, the compensation ions inside the layers or the molecules in interlayer positions can intervene by osmotic process. Lastly, exchange movements between the structural components of clay minerals and those of the skin take place. While penetrating into the tissues positioned close to the hypoderm (Fig. 8.1), these compounds participate in a true therapeutic dynamics, provided that their effects are controlled.

If the application of clay is accompanied by a movement, massage or simple friction, a slight abrasion (or a gumming) is added to the preceding effects, enough to ensure effective cleaning of the treated surface. A further application provides a substantial gain. Furthermore, there are tension lines in the skin that reflect the dominant directions of the collagen fibers. The application of clay on the skin surface is all the more effective and suitable for them because good coordination exists between the directions of these fibers and the exerted forces, which generate the tension forces, as we have just specified.

Pelotherapy is the application on the whole body or just on a part of it of hot marine mud (sometimes with the addition of algae), enveloped or wrapped in an appropriate tissue to slow down the heat loss. The expected effects can be divided into three main categories according to whether they aim at its surface or the near surface of the skin, or whether they aim at an in-depth effect (treatment of contusions, e.g.).

Body hair is completely accessible, inasmuch as it is external to the body, which makes it possible to employ clays for varied treatments ranging from capillary health to that of the scalp.

Obviously these initial properties of natural clay, known as native clay, acquired before its extraction from the clay deposit, can be associated with those of various elements added voluntarily during the final preparation for a therapeutic and commercial aim.

Spas (and also balneotherapy) have developed the use of clay since antiquity. Documentation on this subject is very rich (historic, scientific, medical, or the general public). In order to be used, clay is associated with waters that allow the development of the majority of its properties. Thus, all the elements associated with water become essential: seawater, mineral water, and pure water are much sought after and valued for their respective compositions. With time organic matter additions, often natural (humus, peat, algae, etc.) revealed interesting qualities and today are commonly employed in relation to clay. To be effective, a maturation period is necessary. Altogether this concerns unquestionable knowledge, but where the really effective parameters are sometimes difficult to distinguish among the many properties of clays we described above. These essential actions also cause effective and often complementary secondary actions, such as the relief of tissues then of muscles, which support the circulation or the transport of the biochemical compounds.

8.1.2 *Clay in Pelotherapy: A Review*

Taking into account some inconsistency found in the literature with regard to nomenclature, definition, classification, function, and application of healing clays, particularly in pelotherapy, a review of several basic concepts and of the actual knowledge of the subjects follows.

8.1.2.1 Basic Concepts

Targets, concepts, principles, methods, historical evolution, interactions with other sciences, systems, and implications for society are common fundamentals to any science. Concepts, in particular, can be expressed differently by people from different educational background and experience; also, they can evolve over time. The following definitions try to encompass the actual knowledge of the subject dealt with in the present section.

Clay: A naturally occurring geomaterial composed primarily of fine-grained minerals, the so-called *clay minerals* being fundamental constituents, which is generally plastic at appropriate water content and that hardens when dried or fired. Other minerals, the so-called *nonclay minerals*, and organic matter, could be associated with clay minerals. Also any clay has its own clay fraction for which conventionally some disciplines set a maximum size for clay particles: pedology <2μm; geology, sedimentology and geoengineering <4μm; colloid science <1μm Within the geomaterials, clay is one of the most important for humans due to the great diversity of clay types and applications. As a rule, besides clay minerals clay could contain the so-called associated minerals (silicates such as quartz and feldspar, carbonates such as calcite, oxides and hydrated oxides of iron and aluminium such as hematite, goethite and gibbsite), X-ray amorphous materials such as organic matter, inorganic X-ray quasi-amorphous materials such as opal A and ferrihydrite, and amorphous iron hydroxides. There are several types of clay based on geological, technological and commercial arguments: primary or residual clay, secondary or sedimentary clay, special clay (ball clay, bentonite, kaolin, fuller's earth, fibrous clay and flint or hard clay), common clay (pottery clay, heavy clay), refractory clay, nanoclay, activated clay, etc. Bentonite, kaolin, fibrous clay (sepiolite and palygorskite) and nanoclay are the clay types with more interest for medicinal, pharmaceutical and cosmetic applications. Clay is a major constituent of the so-called *healing clays* and *edible clays*, and is also a major constituent of healing mud or natural peloid and of peloid s.s (stricto sensu) or just peloid, particularly of the so-called *essentially inorganic peloids*, both of them being used for therapeutic and/or cosmetic purposes.

Clay minerals: Fine-grained hydrous phyllosilicates, minerals of planar or non-planar structure that could impart plasticity (if the right amount of water is added) and hardening (if dried or fired) to the clay in which they occur. About seventy species of clay minerals are actually known, species which are distinguished by their crystallochemical specificities and that are usually gathered in groups, such as: serpentine-kaolin; talc-pyrophyllite; smectite, true or flexible mica, vermiculite and chlorite (Guggenheim et al. 2006).

Healing clay or *curative clay*: Clay characterized by healing properties that is used by humans for internal or oral applications in the form of cookies (Fig. 8.2) or clay–potable water dispersion; it could be also used for external or topical applications (*mud therapy* and *peloid therapy*).

Edible clay: A particular type of healing or curative clay used for oral applications and health beneficial purposes (Wilson 2003; Abrahams 2005; Gomes et al. 2008; Ferrell 2008; Kikouama et al. 2009a, b; Kikouama and Baldé 2010; Banenzoue et al. 2014; Gomes, 2017). Edible clay could have a threefold function as a source of mineral micronutrients, protective action against pathogens and toxins (detoxification of noxious or unpalatable compounds present in the diet), and/or antacid action for gastric acidity compensation.

Mud: From a sedimentological point of view mud is a geologic field term for any soft silt–clay mixture with more than 50% of its size fraction smaller than 63 μm and that is plastic when wet. As a popular term mud is also a nonconsolidated

Fig. 8.2 Woman at Port-au-Prince, Haiti, selling cookies made of clay, salt, and animal fat, which are particularly consumed by pregnant women as a dietetic supplement. In Haiti food is scarce and expensive (in *National Geographic*, PT, November 2008, 92:88–91)

fine-grained sticky or slimy sediment of any origin. Mud may also be clay- or carbonate-rich sediment that normally contains some silt-sized components and organic material.

As a healing natural material, mud can be defined as a semisolid mixture of mineral matter and organic matter with mineral water in variable contents. Occasionally mud properties are relevant for healing purposes depending on their origin and composition, and based on these two arguments several typologies and designations can be attributed to mud: *bioglea mud* or *biofilm mud* (called *barégine* in France and *muffe* in Italy), *estuary mud, lagoon mud, freshwater lake mud, gyttja* or *nekron mud, peat mud, river mud, salinas* or *salt pan mud, salt lake mud, sapropel mud, sea* or *marine mud, thermal spring mud*, and *volcanic mud*.

The names of the very singular natural sediments from which healing muds are derived change with the idioms of the countries, often derived from the Latin, as follows.

Argilla, limus, lutum, lamae, caeum (in Latin); *argila, barro, greda, lama, lodo, limo, vasa, silte, sapropel, turfa, biofilme* (in Portuguese); *arcilla, barro, greda, cieno, lama, lodo, fango, limo, vasa, silte, sapropelli, turba, bioglea* (in Spanish); *argile, boue, glaise, fange, liman, limon, silt, vase, sapropeli, tourbe, muffe, mousse, barégine, bioglée, gyttja* (in French); *argille, fango, limi, muffe, torbe* (in Italian); *clay, mud, mire, slime, ooze, loam, silt, sapropel, turf, moor, peat, biofilm* or *bioglea, gyttja* (in English); *ton, schlick, schlamm, torf* (in German); and *tîn, hama, wahal* (in Arabic).

The designation *thermal mud* is attributed to mud whose origin is related to thermal springs.

Mud therapy: The topical use of natural muddy sediment in the form of mud paste for therapeutic and/or cosmetic purposes, mostly in an empiric way and without medical counseling and surveillance; such natural muddy sediment that has been maturated in its geologic site of occurrence is simply called *natural mud* (or just *mud*) or *natural peloid*.

The use of mud for health care purposes goes back to prehistory. Egyptians used muddy sediments from the river Nile, as topical applications to treat disorders of the feminine genital system and to treat skin burns.

Romans were the first to recommend mud baths for therapeutic purposes, mud having origin in deposits occurring at the bottom of seas, lakes, and rivers (such as the Danube) as well as in peat from marshlands.

In France, before Roman occupation, mud baths were practiced in the so-called thermal stations of Dax (Aquitaine), Balaruc-les-Bains (Hérault), and Saint-Amand-les-Eaux (Nord), and after Roman occupation these and other stations were expanded, becoming quite prosperous.

The barbarian invasions in Europe significantly diminished the prosperity of the thermal stations, and only in the Middle Ages did the balneal activity recuperate.

Since the Renaissance (end of the fifteenth century) in Italy thermalism and the use of warm mud have become popular, particularly in resorts located in the Euganean basin famous for thermal springs. Many documents exist reporting the application of warm mud washed out with thermal water when the mud application was finished.

In France what we call today thermal medicine was initiated at the end of the seventeenth century.

In Europe several countries used mud for therapeutic purposes, such as Germany, Italy, Romania, Hungary, Greece, Spain, and Russia, before and after Roman occupation.

Over time and due to public concern related to malpractice in terms of hygiene, the initially usually collective and general baths were replaced by individual baths, and these days the most common treatments involve the use of mud packs, compresses, and facial masks on local and limited regions of the body (lower back, knee, hand, face, etc.).

Certain bacteria pathogenic to humans could be present in muds and in peloids too, and mud packs or cataplasms due to their direct contact with the skin are potentially more hazardous than compresses. Species such as *Clostridium perfringens*, *Clostridium botulinum*, *Clostridium tetani*, *Pseudomonas aeruginosa*, *Staphylococcus aureus*, *Legionella pneumophila*, *Escherichia coli*, and *Candida albicans* are examples of pathogenic bacteria.

Within the European countries, with the exception of certain regions of Germany and France, there are no effective sanitary quality standards or guidelines for muds or peloids utilized for therapeutic purposes. More important than chemical safety is sanitary safety, essential mud microbiological control.

ESPA (European Spas Association), in its General Assembly that was held in 2006, in the Thermal spa or Thermal Resort of São Pedro do Sul, in Portugal, had considered and proposed quality criteria for both medicinal spring water and thermal mud/peloid, but had not quantitatively defined the minimum requirements or

Fig. 8.3 Marly mud paste (seawater is the liquid phase) being applied at Parede beach (on the *left*) and at Porto de Mós beach (on the *right*). Traditionally the mud paste is maintained in contact with the skin until it is fully dried, and afterwards it is washed off with seawater

guidelines, both chemical and microbiological, for these products used in balneology (the study and therapeutic use of medicinal mineral water and of healing mud/medical peloid). In certain regions of countries such as France and Germany (Francois et al. 2005) microbiological specifications of peloids have been established. Outside of Europe, in Cuba, a country where natural peloids and peloids s.s. have a paramount importance for therapeutic and dermocosmetic applications, the sanitary control of these materials has deserved high concern, reason why a regulation was established containing references to mud/peloid chemical and microbiological specifications.

Mud therapy may take place without medical recommendation and surveillance, and a good example of this is the traditional practice that takes place at certain natural sites of the littoral Atlantic of Portugal, more precisely at the beaches of Consolação, Parede, Meco, and Porto de Mós, involving the empiric use of clay pastes made by mixing marly clay (dated from the Upper Jurassic at some sites or the Lower Cretaceous at other sites that occurs in the nearby cliffs) with sweating, for the treatment of rheumatic and cosmetic affections (Gomes 2002; Cardoso-Gomes and Gomes 2015).

Figure 8.3 shows *mud application* at the beach of Parede (near Lisbon), and at the beach of Porto de Mós (in Algarve). Also, mud baths take place in a discontinued salt pan named Barquinha, in Castro Marim (in Algarve), using the *saline mud* deposited at the salt pan bottom.

Peloid: The name "peloid" corresponds to the name *péloïde* proposed in 1933 by Judd Lewis, president of the International Standard Measurements Committee, to encompass all the diverse names of natural sediments to which therapeutic properties were attributed, names naturally varied from country to country, such as: *boue* in France, *fango* in Italy, *schlamm* in Germany, *barro* in Spain, *mud* in English, and *lama* in Portuguese.

The name "peloid" is derived from the Greek word *πξλος*, *pelòs*, meaning "mud," and from the name "peloid", the terms *peloid therapy* and *pelotherapy* were derived.

The peloid designation was definitively adopted at the Conference of the ISMH (International Society of Medical Hydrology) held in Wiesbaden, Germany, in 1938.

Peloid is defined as a: "*natural product (natural sediment) composed of a mixture of mineral water (seawater and lake water included) with organic or inorganic matter that is the result of either geologic or biologic processes or yet of both processes, and is utilized for therapeutic purposes under the form of packs, compresses or baths.*"

This is the official international definition of peloid that was approved in the Fourth Conference of ISMH (International Society of Medical Hydrology) held in 1949, in Dax, France.

Dax is the primary thermal town in France, frequented by about 55,000 patients per year. The famous peloid of Dax (named Terdax®) is produced by mixing the "limon" (clay-rich sediment) extracted from the river Adour with the hyperthermal (39–44°C) calcic and sodic chlorinated–sulphated natural mineral water that emerges from a local spring, and submitting the mixture to maturation inside an hermetic container for 21 days (naturally the original biocenosis existing in the natural sediment is changed with maturation).

About 250,000 tons of peloid are produced every year, distributed by the 16 thermal spas existing in Dax.

In France, in 2005, out of a total of 94 thermal stations there were 73 where peloids were used for therapeutic purposes (Francois et al. 2005).

Not all thermal spas existing in France can mimic Dax in terms of manufacturing specific and differentiated mud on a grand scale. Presently some small spas are buying muds produced and traded by suppliers (as is the case of the firm Argicur®) by mixing their specific thermal waters with commercial pure clay.

The official International Classification of Peloids (the original is shown in Table 8.1) was established and also approved at the same Fourth Conference of ISMH, based on the factors: origin of the solid component, chemical nature and temperature of the mineral water, and maturation conditions.

Before and after 1949, various authors proposed definitions and classifications for peloids, because concepts and classifications can change with time, a situation justified by the progressive scientific evolution. Porlezza (1965) reports the various definitions of peloids proposed in the time interval 1933–1952.

Gomes et al. (2013a) showed the historical evolution of peloid definition and of peloid classification as well. The same authors proposed a new definition of peloid:

> *Peloid is a maturated mud or muddy dispersion with healing and/or cosmetic properties, composed of a complex mixture of fine-grained natural materials of geologic and/or biologic origins, mineral water or sea water, and commonly organic compounds from biological metabolic activity.*

Peloid therapy: The topical use of natural muddy sediment in the form of mud paste for therapeutic and/or cosmetic purposes, under medical counseling and surveillance, sediment that in addition to maturation in the occurrence site has undergone later manipulation and maturation in the artificial environment existing inside

Table 8.1 Classification of peloids

Peloid Denomination	Origin	Mineral Water		Maturation Conditions
		Chemical Nature	Temperature	
Mud[a]	Predominance of the inorganic component	Sulfurous, sulfated, chlorinated, brominated, iodated	Hyperthermal *on spring*	(a) In situ (on the spring source)
			Mesothermal (36–38°C) *on spring*	(b) *On basin*
			Hypothermal *on spring*	
Loam, slime, ooze[b]	*Id.*	Sea, river, or salt lake waters	Hyperthermal *on basin*	In situ
Peat	Predominance of the organic component	Alkaline, carbonated, ferruginous, sulfurous Seawater	Hyperthermal *on spring*	(a) Open container
			Mesothermal *on spring*	(b) Covered container
			Hypothermal *on spring*	
			Hypothermal *on basin*	
Bioglea, biofilm[c]	*Id.*	Sulfurous	Hyperthermal *on spring*	In situ
Bioglea, biofilm[d]	*Id.*	Mineral waters others than sulfurous waters	Hyperthermal *on spring*	*Id.*
			Mesothermal *on spring*	
			Hypothermal *on spring*	
Sapropel	Mixed	Alkaline, ferruginous, sulfurous	Hypothermal *on basin*	*Id.*
Gyttja	*Id.*	Seawater	*Id.*	*Id.*

Adapted from the Original International Classification de Péloïdes, in French, established in Dax, France, in 1949

Note: [a] Predominantly inorganic muddy sediment deposited by mineral spring waters of various chemical and thermal typologies

[b] Predominantly inorganic muddy sediment deposited by sea, salt lake, river, or brackish waters

[c] Predominantly organic muddy sediment (bacteria rich) deposited by sulfurous spring waters

[d] Predominantly organic muddy sediment (bacteria rich) deposited by nonsulfurous spring waters

the spas of thermal resorts and thalassotherapy centers; such artificialized natural muddy sediment is simply called *peloid s.s.* (*stricto sensu*) or just *peloid.*

Pelotherapy: Term of comprehensive meaning used for the topical application of either *natural mud/natural peloid,* or of *peloids s.s.*, with therapeutic or cosmetic purposes (see Figs. 8.4 and 8.5).

Maraver (2006), Gomes et al. (2013a), and Horno (2014) report and discuss, in a synthetic way, relevant information concerned with the historical evolution of pelotherapy.

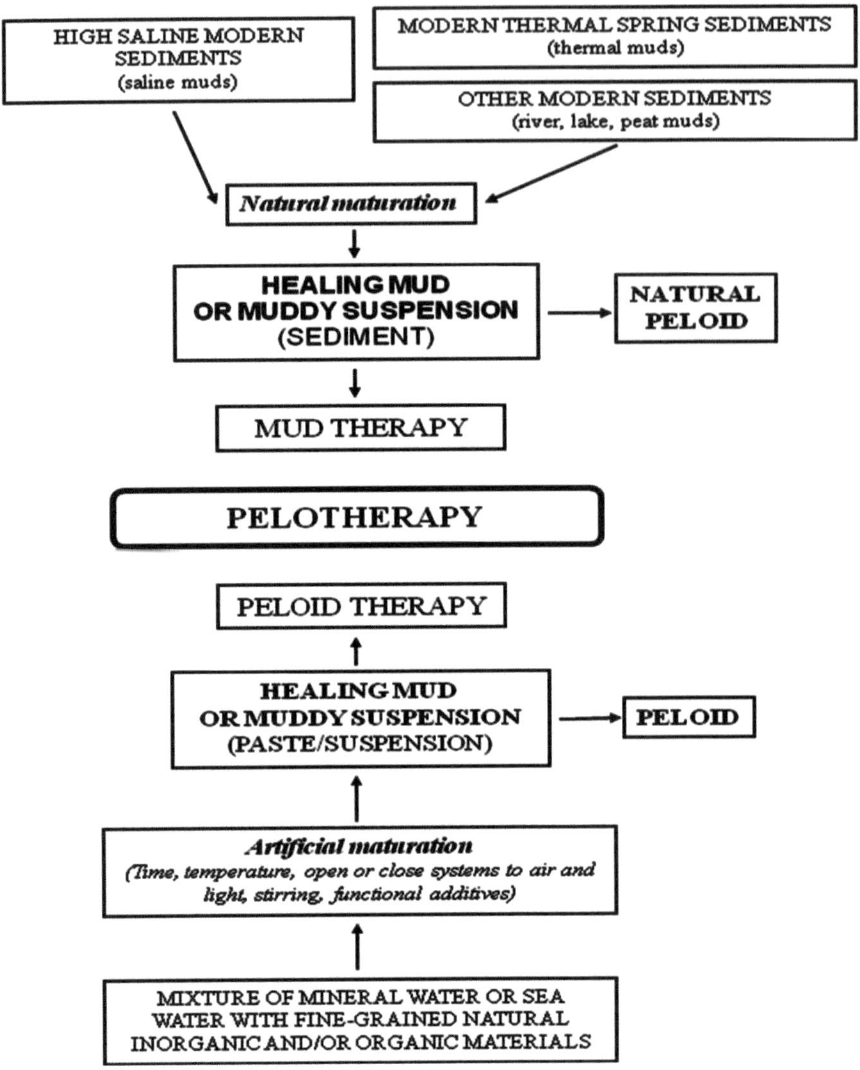

Fig. 8.4 Concepts of mud or natural peloid, mud therapy, peloid, peloid therapy, and pelotherapy. *Source*: Gomes et al. (2013)

Fangotherapy: It is a particular type of pelotherapy that involves the use of *fango* (name of Italian origin attributed to natural mud/natural peloid deposited from thermal springs such as in the volcanic Euganean region in Italy, where important thermal resorts exist, such as Montecatini, Albano, and Battaglia).

Such *natural mud/natural peloid* becomes *artificial peloid or peloid s.s.*, or just *peloid*, after undergoing manipulation and maturation (lasting weeks or months, and needed to achieve the physicochemical equilibrium between peloid solid, liquid, and gas phases) inside the *spas*.

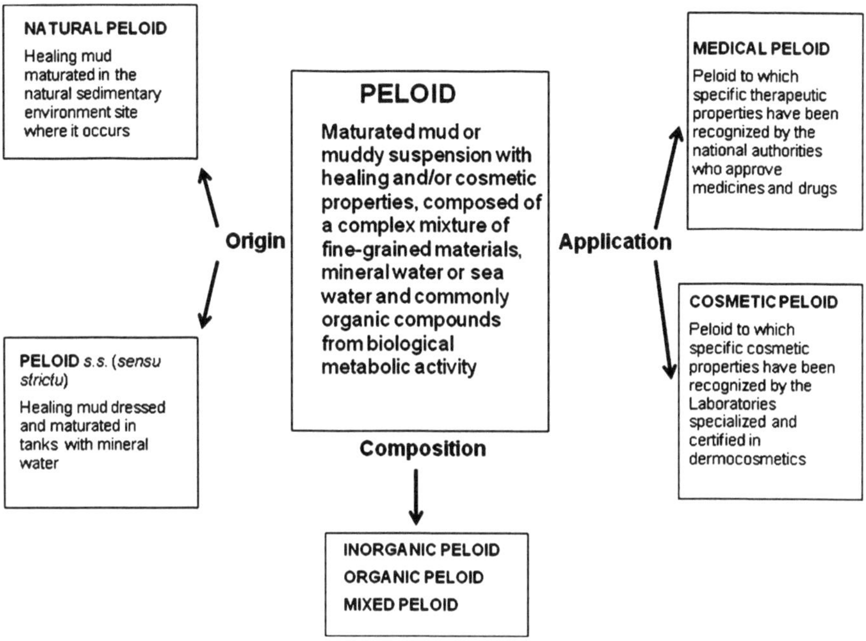

Fig. 8.5 Concepts of natural peloid, peloid *s.s.*, peloid, medical peloid, and cosmetic peloid. *Source*: Gomes et al. (2013)

Another good example of fango/natural mud/natural peloid is the singular natural sulphur-rich warm mud (30–40°C) deposited from hydrothermal springs in the Copahue volcano situated in the province of Neuquén, Argentina. This acidic mud (pH, 2–3) is used in the region for the treatment of both rheumatic and dermatological pathologies. Patients can take mud baths, either in the natural mud pools as in the pool Laguna del Chancho, or undergo similar mud baths but with more comfort, sanitary control, and medical counseling and assistance inside the nearby spa, where the fango/natural mud is refined, beneficiated, and has undergone maturation enabling its designation as artificial peloid or peloid *s.s.* (Ubogui et al. 1998; Baschini et al. 2010). The natural or primary mud is mainly composed of smectite and kaolinite (clay minerals that altogether make up 30–40%) and elemental sulphur (making up 40–50%).

Various studies report that sulphur is one of the most interesting components in thermal muds and thermal waters (Bender et al. 2005; Costantino 2006; Benedetti et al. 2007; Carbajo and Maraver 2015).

Another example of fango, also associated with a similar geothermal system, occurs in Balneário da Coroa, inaugurated in 1811, in Ribeira Grande, São Miguel Island, the Azores (Fig. 8.6). Gomes et al. (2013b) describe the inorganic and organic constituents of this thermal mud or fango.

Natural peloids: In general, they are natural, muddy, and complex sedimentary geomaterials of marine, fluvial, estuarine, lacustrine, or volcanic origin, consisting

Fig. 8.6 Tank containing volcanic natural mud (fango) used in the Balneário da Coroa, São Miguel Island, the Azores

of mixtures of solid, liquid, and gas phases, each containing diverse and specific components.

Based on origin, natural peloids can be classified in two classes: primary peloids or virgin peloids (if the solid component has been mechanically transported as particulate dispersed material and deposited in the mineral water from a spring), and secondary peloids (if the solid component and the mineral water come from different sources; Veniale 1998). Based on global composition, more precisely on the quantitative ratio of inorganic–organic components, natural peloids can be classified into three groups: essentially inorganic peloids (mud of fluvial, lacustrine, or marine origin and fango of volcanic origin); essentially organic peloids (peat and biofilm); and mixed inorganic–organic peloids (*sapropel* and *gyttja*).

Also based on the mineral and chemical composition, particularly of the solid component, natural peloids can be classified into three groups: phyllo-peloids (characterized by significant content of phyllosilicates represented by clay minerals), organo-peloids (characterized by significant participation of organic matter), and sulfo-peloids (characterized by significant participation of sulphur existing both in the inorganic and organic matter as well as in mineral waters, under the forms of S, H_2S, or metal sulfides) (Pozo et al. 2010).

Also, natural peloids' liquid component could be diverse in terms of mineral water chemistry: *chlorinated* (saline, ≥1 g/L total dissolved solids, Na^+—dominant cation, Cl^-—the dominant anion); *bicarbonated* (≥1 g/L total dissolved solids, Ca^{2+} and Mg^{2+}—dominant cations, HCO_3^-—the dominant anion); *sulfated* (≥1 g/L total dissolved solids, SO_4^{2-}—dominant anion); *sulfuric* (total sulfur ≥1 mg/L of HS^-, or $S_2O_3^{2-}$, or S^{2-}, or H_2S); *iodated–brominated* ≥1 mg/L of I^-or >5 mg/L of Br^-; *ferrous* (≥10 mg/L of Fe^{2+} or Fe^{3+}). In terms of temperature measured at their emergence, mineral waters can be classified into thermal (≥25°C) or not thermal. All mineral waters, and spa or medicinal waters being used in balneology, in particular, bear a small dissolved fraction of organic compounds that could be recognized and identified by gas chromatography analyses (Varga 2010).

Based upon their application and function natural peloids can be classified into two groups: therapeutic or medical peloids and cosmetic or skin care peloids (Gomes et al. 2013a). The classifications referred to could be applied as well as to artificial peloids.

Natural peloids and peloids *s.s.* (*stricto sensu*) also require a number of suitable properties for use in pelotherapy, such as: consistency, viscosity, adhesiveness, moisture content, heat capacity, cooling rate, exchange capacity, spreadability, abrasiveness, ease of handling, and pleasant sensation when applied to the skin (Veniale 1997). Within these properties the peloid's thermal and rheological properties have deserved particular interest and study (Rambaud et al. 1986; Ferrand and Yvon 2000; Yvon and Ferrand 1996; Armijo 1991, 2015; Casas et al. 2013).

Mud therapy based on mud or natural peloids (maturated and applied at or near the natural site of occurrence) is progressively losing applied interest in favor of peloid therapy based on artificial peloids or peloids *s.s.* (*stricto sensu*) manipulated, maturated, and applied inside spas, due to the following facts.

1. The access to natural peloids is becoming more and more difficult due to environmental restrictions (the deposits are, as a rule, located inside environmentally protected areas).
2. The geologic sites where natural peloids occur are open systems highly vulnerable to anthropogenic contamination, particularly of pathogen microorganisms, making control of their sanitary safety infeasible.
3. The use of natural peloids could be hazardous, as happens with mud deposited by acidic volcanic waters characterized by very acid pH (2–3) and usually bearing high heavy metal content.
4. The complexity of natural peloids' composition and characteristics makes difficult not only the identification of the healing active principles, but also understanding the healing mechanism of action and process, and the control of both composition and sanitary state as well (related to the presence of potentially hazardous chemical elements and pathogen microorganisms).

As a matter of fact the aforesaid facts justify the increasing interest for peloid therapy based on *manipulated peloids*, that is, peloids *s.s.*, particularly those called designed and engineered peloids.

Designed and engineered peloids: They are *peloids s.s.* that can be produced, either inside specialized manufacture units, or inside the spa facilities where they are applied. Designed and engineered peloids can be easily manufactured using, for instance, specific mixtures of one essentially monomineral and commercial clay (e.g., kaolin or bentonite), preferably of pharmaceutical grade, with one specific mineral water (spring thermal water or seawater), that after undergoing manipulation and maturation could be beneficiated (for instance, through the incorporation of *healing-oriented functionally active additives*, either natural or synthetic, characterized

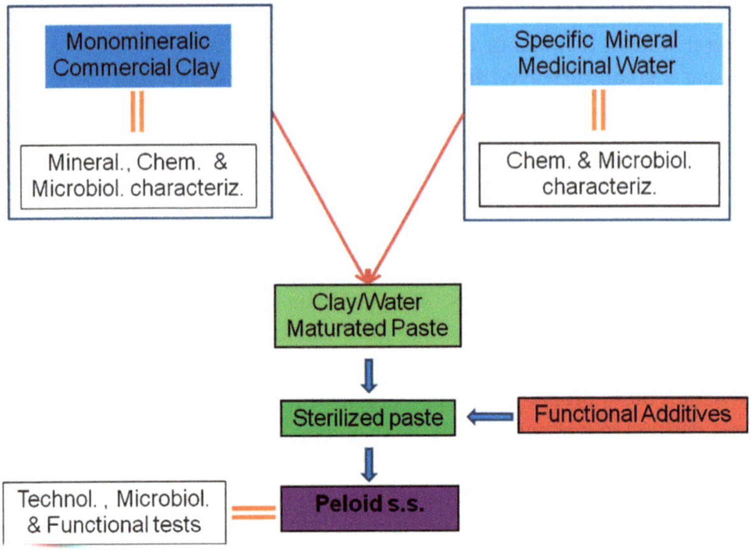

Fig. 8.7 Flowchart of the preparation of *designed and engineered peloids* (Gomes 2015b)

by analgesic, anti-inflammatory, antioxidation, anticellulite, antipsoriasis, or anti-acne, among other properties).

Figure 8.7 shows the flowchart of the preparation of designed and engineered peloids. Due to their simple composition and processing, designed and engineered peloids of both medical and cosmetic typologies can easily be periodically submitted to both composition control (identification and quantification of inorganic or mineral and of organic constituents) and sanitary control (identification and quantification of pathogenic microorganisms)

In summary the advantages of using a designed and engineered peloid are fourfold:

1. Previous and fair knowledge of the properties of peloid individual components, one specific monomineral clay, and one specific mineral water. The specific monomineral clay could be selected, either by its almost chemical inertia (for instance, kaolin which has a very simple chemical composition and structure characterized by electric charge z, globally negative but almost null), or by its high chemical activity (for instance, bentonite which has a more complex chemical composition and structure characterized by significant globally negative electric charge z = 0.2–0.6. The specific mineral water could be selected within medium to highly mineralized waters, such as thermal spring water or seawater).
2. Simple and brief manipulation (enough to homogenize the mixture as much as possible (using a mechanical stirrer) and eventually to allow the introduction and dispersion into the paste clay/mineral water of, either pharmacologically and medically recognized functional active principles (natural or synthetic, and inorganic

or organic), or organic compounds that could promote the enhancement of skin permeation allowing overcoming the *stratum corneum* barrier.
3. Relatively brief maturation (1–2 weeks).
4. Simple and effective quality control, both chemical and sanitary, in accordance with the guidelines proposed by AFTh (Association Française de Thermalisme; in Fouquey 2014).

According to Gomes et al. (2015a) clay and mineral water are the effective carriers or vehicles as well as fixers and suppliers of either natural or synthetic active principles responsible for peloids' therapeutic or cosmetic properties. Based on this and characterizing designed and engineered peloids it will be possible to design and develop the so-called *à la carte peloids*, genuine and differentiating, of any specific thermal resort or thalassotherapy center, natural mineral water or seawater being, respectively, the peloid liquid phase. The authors also consider that the peloid activity could depend more on the chemical particulars of both specific natural mineral water and of medically/pharmacologically recognized specific functional additives than on the physicochemical and chemical particulars of the clay that would act mainly as a heat conservation and transfer agent. If it is true that the peloid liquid phase is more important than the peloid solid phase, then clay plays a threefold important role: it can remain warm for a longer time than the semisolid paste that is in contact with skin, it can modify the properties of the liquid phase through the passage into solution of the water-soluble compounds existing in the solid phase and of the reversibly fixed chemical elements existing at the surfaces of the fine-grained particles of both clay minerals and organic matter (humus), and it can facilitate and extend for a longer time the interaction and, eventually the percutaneous absorption of the therapeutic or cosmetic active substances favored by the heat previously provided to the peloid (Gomes 2015b).

With regard to the time required to achieve peloid optimum maturation Fernández-Gonzalez et al. (2013), based on a study that involves one artificial peloid prepared by blending one of two clays (kaolin and bentonite) with one mineral medicinal water, found that 1 month of maturation time would be sufficient to provide what is called the "optimum peloid." Traditionally for peloids of higher complex composition 6–12 months of maturation are required to attain the optimum maturation state.

Also with regard to the optimum peloid prepared with the specific mineral medicinal water used in the balneary of a certain thermal resort, Carretero (2013) recommends a methodology contemplating the following main steps:

1. Characterization of both solid (specific clay) and liquid phase (specific mineral medicinal water)
2. Mixture of both solid and liquid phases
3. Maturation of the mixture
4. Characterization of the obtained peloid
5. Study of therapeutic and/or cosmetic efficacy after peloid application in patients
6. Global evaluation of the studies being carried out, contemplating the determination of the clay showing the best performance, as well as the determination of the

best mode of making the mixture and peloid maturation conditions, and of peloid optimum maturation time.

Extemporaneous peloid: Healing mud prepared locally and at the time of use, blending one common or special clay (kaolin or bentonite) or algae with mineral water or seawater without undergoing maturation; it is not a real peloid *s.s.* (*stricto sensu*) because maturation does not occur.

Meijide et al. (2010) and Arribas et al. (2010a) studied and assessed the effects on the mechanical and functional characteristics of skin from the application of extemporaneous peloids based on bentonite mixed with both sulphurous thermal water and seawater, as well as the clinical effects (expressed by scaling off and inflammation) on psoriasis.

Paramud or *parapeloid* or *parafango*: Healing mud that is the product of the blend, for instance, of bentonitic or illitic clay and paraffin; as a matter of fact parapeloid or parafango is not a type of peloid; it is a peloid succedaneum. Parapeloid or parafango is only applied as cataplasm and always hot (not above 50°C) in physiotherapy.

8.1.2.2 Peloid Action, Function, and Benefits

As a rule, both *natural peloids* and *peloids s.s.* are three-phase systems consisting of: one solid phase (inorganic, organic, or mixed inorganic/organic), one liquid phase (thermal spring, river, lake, or seawater), and one gas phase (H_2S, CO_2, NH_4, CH_4, Rn).

As a rule, and in quantitative terms, the gas phase is irrelevant, but it could be therapeutically relevant. According to Benedetti et al. (2007), Burguera et al. (2012), and Fioravanti et al. (2013) H_2S, either formed from organic matter decomposition (a situation common in the mud of salt pans), or existing in solution in sulphurous water (frequent in thermal mineral waters), would act as an anti-inflammatory and antioxidant agent in human osteoarthritic articular chondrocytes. According to Carbajo & Maraver (2015) H_2S and its derivates have remarkable influence on the biomechanical and physiological properties of skin, being responsible for arterial vasodilatation. The authors report that H_2S and NO are considered gasotransmitters molecules in multiple physiological and physiopathological functions, and have potent cytoprotecting properties. In balneology H_2S could be absorbed, either through inhalation or through the skin (bathing in water or applying peloids, both bearing the molecule).

Peloids' inorganic component is mainly composed of a fine-grained sediment called clay, itself a very complex system that consists of clay minerals (as a rule, three or four species) and of nonclay minerals (as a rule, three or four species), sometimes bearing organic substances and associated noncrystalline compounds.

To study the clay component, mineralogy, geochemistry, physics, and physicochemistry methods are currently being employed. Properties such as clay total content, clay minerals, and nonclay minerals' identification and quantification, organic matter total content, grain size distribution, specific surface area, adsorption capacity, cation exchange capacity, exchangeable cations, and pH have been the most studied properties thus far, despite that the clay content could be merely residual, as happens for instance in some of the so-called essentially organic peloids, such as peat, sapropel, and biofilm. As a matter of fact over the last 20 years, the peloid inorganic solid component in particular has been thoroughly studied by specialists, geologists, physicists, and chemists.

The articles by Carretero et al. (2014) and Pozo (2015) are very recent examples of this kind of study. The authors thoroughly studied three Spanish commercial clays, and used them mixed with distilled water to make artificial peloids whose physical and physicochemical properties were compared with equivalent properties of Spanish and Turkish natural peloids. The authors, having used distilled water, apparently devaluated the liquid component relative to the solid component.

On the other hand, studies involving the peloid organic solid component are very limited (Odabasi et al. 2007; Tserenpil et al. 2010; Suárez et al. 2011; and Gomes et al. 2013a, b). Nowadays, it is clearly understood that a variety of compounds can be found in peloids, naturally in higher number in the essentially organic peloids. GC-MS and similar methods are frequently used to identify polar and nonpolar organic compounds. Within those compounds some could be biologically active (antioxidant, anti-inflammatory, and immune stimulatory), such as humic compounds, carboxylic acids, terpenoids, steroids, fatty carboxylic acids, and hydrocarbons.

In the semisolid paste being formed mixing and homogenizing both solid and liquid phases, *peloid thermal properties*, such as heat capacity, specific heat, heat diffusiveness, and cooling rate, as well as *peloid rheological properties* (plasticity and viscosity), have also been thoroughly studied. The article by Pozo et al. (2013) reporting the study of thermal and nonthermal physical and physicochemical properties of five Spanish peloids is a very representative example of the importance of peloid thermal properties. Any short or long-lasting interaction between the solid and liquid component can promote, particularly at the chemical level, mutual changes in both of them, changes that could be reduced and controlled when the designed and engineered peloid is being used. The great complex composition of both natural peloids and peloids *s.s*, does not facilitate at all the identification of the beneficial or healing agents or factors existing in a certain peloid.

Clinical evidence based on epidemiological studies, and available in the specialized literature, emphasizes the benefits of mud therapy and peloid therapy on degenerative and inflammatory rheumatism taking advantage of peloids' analgesic, anti-inflammatory, antioxidant, and antimicrobial actions. Several authors have emphasized the benefits of peloid properties and peloid therapeutic functions, including Veniale (1996, 1997, 1999), Carretero (2002), Cozzi et al. (2004), Codish et al. (2005), Carretero et al. (2006, 2013), Cantarini et al. (2007), Maraver (2013), Tricás et al. (2014), Roques (2015), Fioravanti and Chelesschi (2015), Meijide et al. (2014, 2015), and Gomes (2015b).

With regard to peloid benefits on rheumatic diseases (particularly, hand osteoarthritis, knee osteoarthritis, and lower back pain) recent works (Bellometti et al. 1997,

2000, 2002, 2005, 2007; Flusser et al. 2002; Vaht et al. 2008; Odabasi et al. 2009; Mahboob et al. 2009; Ceccarelli et al. 2010; Fraioli et al. 2010; Fioravanti et al. 2010, 2011, 2014; Sarsan et al. 2012; Liu et al. 2013; Espejo-Antúnez et al. 2013; Tefner 2014; Meijide et al. 2015; Tenti et al. 2014) constitute very important benchmarkers, inasmuch as they consider that the mechanisms of action in rheumatic diseases, either through bathing in mineral medicinal water, or through mud bathing and mudpack application, are still not fully understood, in spite of the health benefits that have been achieved.

For Fioravanti et al. (2011) the chemical effects of both mudpack and mineral medicinal water on pain relief were less clear than the physical effects (temperature and hydrostatic pressure of water on the skin).

Although much research has been done thus far, much experimental research is still required in order to appraise and confirm the effective mechanisms of action and the health benefits of mud/peloid applications. Experience has shown that in the case of mud/peloid packs utilized in the treatment of musculoskeletal disorders (e.g., knee osteoarthritis), randomized, controlled, double-blind studies are always recommended.

Also, there is evidence of the positive effects of peloids on dermatological affections, especially on psoriasis, acne, and seborrhea, as well as on skin care functions (cleansing, degreasing, exfoliating, hydrating, tonifying, and firming). Several authors have emphasized the benefits of peloid properties and peloid dermatological and aesthetic functions (Ubogui et al. 1991; Panico and Imperato 2009; Matz et al. 2003; Arribas et al. 2012; Baschini et al. 2014; Meijide et al. 2010, 2014; Garcia 2014).

Skin care products also include products intended to clean, perfume, change appearance, correct body odors, and protect against solar radiation and other aggressive environmental factors. The formulations of such products (pastes, ointments, creams, gels, and foams) involving natural constituents (certain minerals and clay minerals, in particular) and synthetic constituents are designed to respond to specific functions.

Barros et al. (2010) and Pena-Ferreira et al. (2010a, 2010b, 2011) have shown the effectiveness of several formulations (similar to those common in peloids) involving a clay component rich in smectite, either with cleansing and clarifying effects in facial masks, or with anticellutic effects. Also, Meijide et al. (2010) and Arribas et al. (2010a, b), respectively, have studied and assessed the effects on the mechanical and functional characteristics of skin of the application of extemporaneous peloids based on bentonite mixed with both sulphurous mineral medicinal water and seawater, and the clinical effects (expressed by scaling off and inflammation) on psoriasis of the thermal natural peloid from La Toja (Spain).

In regard to the use of natural peloids in cosmetics there has been renewed interest in facial/body masks based on clayey materials. In general masks are defined as cosmetic products to be applied on both face and body in more or less thick layers for a limited period of time (Khanna and Gupt 2002).

Clayey masks are considered the oldest cosmetic preparations used in beauty treatments (Denaverre 1975; Baby et al. 2004). Specific clays, in addition to being used as excipients for their emulsifying, gelifying, and consistency, are used as active substances in cosmetic preparations (Zague et al. 2007). Within the various effects attributed to facial/body masks the following actions should be enhanced: cleansing, exfoliating, tonifying, degreasing, astringency, hydrating, tensoring, and agreeable sensation to touch (Wilkinson and Moore 1982).

Finally there are references to health benefits of the application of peloids in gynecological affections (Artymuk et al. 2010; Beer et al. 2013; Rondanelli et al. 2012) and other types of cutaneous or subcutaneous affections (Silva et al. 2015).

8.1.2.3 Clay/Clay Minerals and Human Health

Both therapeutic and cosmetic uses of clay and clay minerals (hydrous phyllosilicates) have been reported by several authors such as Bech (1987), Viseras and López-Galindo (1999), Veniale (1999), Reinbacher (2003), Gomes and Silva (2007, 2008), Carretero (2002), Carretero et al. (2006, 2013), Viseras et al. (2007), Rautureau et al. (2010), De Vos (2010), Young (2011), Gomes (2013); Gomes et al. (2013a), Meijide et al. (2015), Williams and Hillier (2014), and Gomes et al. (2015a).

Clays and clay minerals such as kaolin, bentonite, attapulgite (mineralogically discredited name, abandoned in favor of the name palygorskite), sepiolite, and talc are reported in pharmacopeias (for instance, the European Pharmacopoeia (EP) and United States Pharmacopeia (USP)). Being considered medicines they could participate in pharmaceutical formulations as active principles and/or as excipients (Viseras and López-Galindo 1999; Lopez-Galindo and Viseras 2004; Lopez-Galindo et al. 2007; Viseras et al. 2007, 2010). In particular the use of bentonite and kaolin in human health has been the object of significant research (Novelli 1996, 1998, 2000; Cara et al. 2000; Viseras et al. 2006; Williams et al. 2009, 2014; Casás et al. 2011; Carretero et al. 2013).

Clays and clay minerals may enter the human body by inhalation, ingestion, or dermal absorption.

The main properties of clays (attributed to clay minerals) that justify the use for healing purposes of the so-called healing clays and edible clays are as follows:

1. Ubitiquous geological occurrence at or near the earth's surface; therefore clay is easily available for access, extraction, and use.
2. Clay minerals very small particle size (as a rule less than 2 μm, and that could goes down to nanosize), and diverse and mostly anisometric particle shape (platy in most clay minerals, but fibrous, tubular and spherical in others); both characteristics being essential conditions to yield clay high specific surface area (SSA), property that is a determining factor when clay is used, for instance, either for the coating of stomach and small intestine in order to alleviate nausea and indigestion, or for topical coating and adhesiveness of/to skin, or yet for drug delivery; bad characteristics when clay could provide, for instance, toxicity associated to long-term inhalation.
3. Low hardness and abrasiveness provide an agreeable sensation when the clay paste is in contact with the skin, and is spread on it.

4. Great affinity to water facilitates the formation of plastic pastes showing adequate flowing, spreading on, and adhesiveness to the skin.
5. Particle surface electric charge globally negative mainly due to structural point defects (atomic substitution, atomic vacancies, particle edges broken bonds) existing in clay minerals; pH dependant negative electric charge that could be modified (even compensated and neutralized) by particle superficial coatings of naturally associated quasi-amorphous or amorphous electropositive iron hydroxides; the value of clay mineral layer charge depends on clay mineral nature and specificity; in clay minerals the net negative layer charge per half unit cell varies within the range 0–1, for instance, clay minerals of kaolinite group have a low net negative layer charge (0.1–0.2) per half unit cell, whereas clay minerals of smectite group have an intermediate layer charge (0.2–0.6) that is responsible for layer expansion and shrinkage in case the mineral is in either wet or dry environment and conditions respectively, and clay minerals of illite group have an high layer charge (0.6–0.9), reason why they do not expand when hydrated; the globally negative electric charge of clay minerals being dependant of the pH of the environment of occurrence could be modified (even compensated and neutralized) by particle superficial coatings of naturally associated quasi-amorphous or amorphous electropositive iron hydroxides; such modifications will have consequences on clay relevant properties such as ion exchange, plasticity and rheology.
6. High adsorption/absorption capacities allowing the adsorption/absorption of toxins, bacteria and viruses, and their elimination from the skin, making efficient the use of certain types of mud/peloid in the treatment of cutaneous affections such as, acne, seborrhoea, eczema, psoriasis, and allowing too the use of certain nanotubular clays, as is the case of halloysite, in the process of controlled release of drug delivery.
7. Clays of alkaline character can buffer the gastrointestinal acidic pH, whereas clays characterized by acid pH, if similar to the skin pH, avoid skin irritations.
8. High cation and anion exchange capacities promoting the reversible fixation and the exchange of chemical elements/compounds (polar in particular), for instance, either between the clay making facial masks and the skin, or the incorporation of the elements/ with medicinal interest.
9. Exchangeable reduced elements, such as Fe and its oxidation, are responsible for the bactericide character of certain clays including some classified as French green clays , that can become toxic to human pathogens.
10. High capacity of heat retention and low heat diffusiveness, important characteristics for mud pack or peloid pack essentially based on clay minerals since they can be applied warm (not above 50 °C) or cool, depending on the nature of the affections to be treated; mud or peloid heat loss rate or cooling rate (as a rule, 15–20 min from 50 °C down to 37 °C) being conditioned by factors such as: (1) clay global composition and texture; (2) nature and amount of clay minerals; (3) nature and amount of associated organic substances; (4) and nature of the liquid phase (spring water, sea water, lake water…) whose amount being also a very influent factor, as a rule, is within the range 40–80 %.
11. Great diversity of natural colours.

In the preparation of medicines incorporating clays and clay minerals these must fulfill certain requirements comprising control of the following properties: chemical stability and inertia, chemical safety, microbiological safety, textural adequacy (particle size, particle size distribution, and particle aggregation), water content, and color.

Clays can enter in pharmaceutical formulations as suspending and stabilizing agents, whose functionality depends a great deal on particle morphology and the surface electric charge of clay minerals.

In pharmacy clay and clay minerals can play important roles in modified drug delivery systems, allowing the optimization of time, rate and site of drug release. However some systems could exist in which drug/clay interaction can affect the bioavailability of the drug. Nanometer size clay particles can be incorporated into a polymeric host carrier in order to control de diffusion rate of a dispersed slow-released drug.

Synthetic clays such as those called "anionic clays" or "lamellar double hydroxides (LDHs)", composites clay-polymers and nanocomposites clay-polymers are being used to slow/extend or vectorize the release of drugs, and consequently to increase their bioavailability.

Relevant research has been carried out on these subjects in recent years, and important information and experimentation has been provided (Aguzzi et al. 2007; Buzea et al. 2007; Lvov et al. 2009; Veerabadran et al. 2009; joshi et al. 2009; Viseras et al. 2010; Suresh et al. 2010; Vergaro et al. 2010, 2012; Rodrigues et al. 2013).

Kaolinitic clays, illitic clays, and smectitic clays are the most used *healing clays* and *edible clays* for therapeutic and cosmetic purposes.

Kaolins are being considered effective as hemostatic wound dressings, and various forms of kaolin have been shown to be antibacterial, and are also being utilized in drug delivery (Williams and Hillier 2014). Kaolinite (shows platy morphology) and halloysite (can show either platy or tubular or even spherical morphologies) are the main clay minerals that could be present in kaolins. Compared to platy kaolinite forms, halloysite tubular forms have higher specific surface areas, and consequently greater potential to bind, in particular, cationic drugs, enabling the conditioning and control of their release rate. Halloysite nanotubes (with c. 50–70 nm of external diameter and c. 15 nm diameter lumen) are able to entrap and carry inside the nanotubes' lumen drug molecules that could be used in biomedical applications.

Smectite clays are characterized by much higher surface areas and unit layer charges compared to kaolin's and have interlayer-free and variable size spaces where drug molecules could be entrapped or encapsulated enabling their biomedical use as drug delivery systems. Within the clay mineral species constituting the smectite group, montmorillonite has been the most used species.

Also, the bactericide character has been reported in the case of the so-called French green clays (Williams and Hillier 2004, 2008, 2011; Williams and Haydel 2010; Otto and Haydel 2013).

Smecta® and Beidelix® are two drugs commercialized in pharmacies, whose active substances are the clay minerals diosmectite (dioctahedral smectite of Al–Mg montmorillonite typology) and beidellite (also dioctahedral smectite but richer in Al), respectively. Both drugs are ingested in the form of powders after being dispersed in potable water.

8.1.2.4 Healing Clays: Mechanisms of Action

The mechanisms of action of peloids are still far from being well identified and understood, despite peloid application and healing action being recognized and recommended for the last 75 years by the ISMH (International Society of Medical Hydrology).

There are authors, such as Hattori (1963), Rambaud (1989), Serofilli (1980, 1994), San Martin Bacaicoa (1994), Veniale (1996), Roques (2004), Carretero and Pozo (2007), De Michele et al. (2006), Gomes and Rautureau (2013), Legido and Mourelle (2008), Maraver (2008), and Rautureau et al. (2010) who have tried to explain the close relationships existing between pelotherapy and health care.

Particularly over the last 20 years several authors have tried to find scientific explanations for the physiological benefits of pelotherapy that could be summarized as: (1) analgesic action, (2) anti-inflammatory action, (3) antioxidant action, (4) diminishing chondrolytic activity, (5) contraction of both plain and straight muscular fibers, and (6) antimicrobial action.

Recently Maraver (2013) and Maraver et al. (2015) produced excellent reviews on the state of the art regarding the mechanisms of peloid actions, and classified them into two main categories: thermotherapeutic and chemotherapeutic.

A. *Thermotherapeutic action*: Thus far it is the main action attributed to mud/peloid, in consequence of the temperature of their application; as a rule, they are applied directly onto the skin in layers 1–2 cm thick normally for periods of 20–30 min after being warmed up to 42–45°C; in consequence of the heat, local effects occur such as vasodilatation and hyperemia that improve blood irrigation and circulation in the skin and underlying tissues, and general effects such as liberation of histamine and acetylcoline and modification of serum levels of amino acids (Pastor 1998; Fernández-Lao et al. 2012; Sarsan et al. 2012; Maraver 2013; Meijide et al. 2015).

B. *Chemotherapeutic action*: Numerous studies have demonstrated that the anti-inflammatory, analgesic, chondroprotecting, and immunological actions could be attributed to mud/peloid chemical properties (Carretero 2002; Carretero and Pozo 2007; Gomes and Silva 2007) and to the absorption of minerals (Tateo et al. 2009; Carretero and Pozo 2010) and organic substances (Teixeira et al. 1996; Curri et al. 1997; Tolomio et al. 1999; Nissenbaum et al. 2002; Torrella 2006; Odabasi et al. 2007; Tserenpil et al. 2010; Kim et al. 2010; Suárez et al. 2011), most resulting from the maturation process (Galzigna et al. 1996; Galzigna et al. 1999a; Galzigna et al. 1999b; Sanchez et al. 2002; Veniale et al. 2004; Carretero and Pozo 2007, 2008; Gámiz et al. 2008, 2009; Fernández-Gonzalez et al. 2013).

Mud/peloid analgesic action, anti-inflammatory action, and cicatrizing action (San Martin Bacaicoa 1994; Roques 2004; Bagnato 2004; Meijide et al. 2015), as well the immunological action through the decrease of NO levels in serum or through the reduction of oxygen reactive species (ROS; Fioravanti 2011, 2015) have been enhanced by these authors. The transdermal absorption of inorganic (minerals) and/or organic molecules existing in the mud/peloid could be responsible for both anti-inflammatory and chondroprotecting action (Carretero et al. 2010; Tateo and Summa 2007, Tateo et al. 2009; Beer et al. 2003, 2013).

Peloids are used under *medical counseling and accompaniment* in Health Resort Medicine Centers or in Spa Therapy Centers. The health benefits of peloids are being attributed to a combination of effects, and within these, the thermal and chemical effects are currently the most emphasized.

There is sufficient clinical evidence in the specialized literature confirming, at least, the short-term benefits of peloids; these benefits are attributed to a combination of those factors referred to above, despite both thermal and chemical factors being considered the most effective. Thermal factors are generally accepted and attributed to the solid phase (clay), whereas there is little evidence relative to the second factor, although one might expect that minerals or organic substances dissolved in the liquid phase (mineral water) of the peloid could interact and provide dermal absorption during the peloid application.

On the one hand, the solid phase of the peloid paste favors the maintenance of peloid heat along peloid application as well as the contact and eventual dermal absorption of chemical elements and molecules reversibly fixed at clay minerals' surfaces by ion exchange. On the other hand, minerals and organic substances dissolved in the liquid phase of the peloid paste become available for dermal absorption during the entire time (around 20 min) of peloid application.

However, the hydrophobicity of the skin barrier formed by the epidermis *stratum corneum* layer does not facilitate the percutaneous penetration of hydrophilic substances, minerals, and polar organic compounds (Gomes et al. 2015a). Mineral salts, trace elements, and organic polar compounds carried in solution in the hydrophilic mineral water cannot easily penetrate the skin (Bolzinger et al. 2012).

Recent and very interesting research is looking at transdermal delivery systems attractive for both topical and systemic therapeutics (Schaefer and Redelmeier 1996; Schnetz and Fartasch 2002; Beer et al. 2003; Godin and Touitou 2007; Steinstand 2012; Gupta et al. 2013). The so-called skin permeation enhancers (surfactants, natural oils, hydrocarbons, fatty acids, among others) that are being looked at by several researchers can improve the penetration of bioessential substances by perturbing the barrier function of the *stratum corneum* (Naik et al. 1995; Tanojo et al. 1999; Lim et al. 2014; Pandey et al. 2014).

In all cases, to favor skin permeation, immediately before peloid application, skin cleansing, exfoliation, and hydration are recommended.

According to Araújo et al. (2015), to overcome this hindrance, the cosmeceutical/cosmetics industry is innovating and developing studies involving nanobiotechnology to create and encapsulate effective functional molecules in nanocarrier systems, namely liposomes, polymeric and solid lipid nanoparticles, and nanoemulsions, that are considered fundamental to increase skin permeation/penetration to minerals and organics carried in solution in mineral water. However, the peloid solid substances can modify the properties of the liquid phase through the passage into solution of the water-soluble compounds existing in those substances and of the reversibly fixed chemical elements at the surfaces of the fine-grained particles of both clay minerals and organic matter.

The simple composition and easy manipulation of designed and engineered peloids, and also the easy incorporation of amphiphilic organic compounds (containing polar water-soluble chemical groups and water-insoluble chemical groups) could overcome such hindrance.

8.2 Internal Applications

8.2.1 Food Supplement

As a food supplement clay consumption is an established fact in all civilizations. More generally *geophagia* is literally defined as "soil or earth eating," but in this work the concept is limited to clay consumption.

The nature and assimilation of food plays a central role in the general equilibrium of the body. Clays have always been used in more or less large quantities and absorbed at shorter and shorter intervals in the food process with which they are associated. From this point of view there is often confusion between the concepts of food and drug. (It is necessary to pay great attention to this distinction.)

The act of geophagia relates to consumption and must be considered separately or complementary to geotherapy which relates to consumption with a therapeutic aim. Clays, the so-called edible clays, have also been used to satisfy famished populations by ensuring, without much danger, an alleviating and hydrating alimentary bolus of a sufficient volume. Nevertheless, some types of clay can facilitate food digestion when associated with it.

Let us recall, on a purely historical level, that one of the current methods of recognizing the clay character of a mineral sample is its property "to adhere to the tongue," thus recognizing the aptitude of a sample to fix water, a property specific to many types of clay. By this affinity this simple sign indicates the recognition of clay by the tongue, the first organ of the digestive chain.

Anthropologists found evidence of the use of clay in the food of very old civilizations. They relate to the practice of geophagia: "clayey soils can be eaten in the raw state with some seasonings or after being cooked." One may consult the book: *The Treaty of Simples*, a translation by L. Leclerc, devoted to the study of translations from Greek authors by Arabs, which has precise details on the subject. The markedly green soil of Sîrâf (probably nontronite clay), smoked with orange peel becomes red and tastes good. The author adds that it is seldom eaten smoked. This book also quotes Ali Ibn Mohamed in connection with the soil of Nisabour which is eaten "well roasted" or equally raw.

Judiciously selected clay can be used for its property to form a gel in order to adjust the texture of dishes or condiments during their preparation. It has been used this way for the preparation of fatty emulsions such as mustards or margarines.

In some modes of food cooking, clay is used to form a shell that hardens under the action of heat and, in the course of cooking, protects the food from external contacts. The cooking of fish and poultry can thus benefit from preparation facilities, protection during cooking without contact with fire, and of conservation of flavors thanks to the formation of a tight protective shell. This method enables avoiding thermal cracking of many food molecules. Thus, the writer Colette in her book *Prisons et Paradis* mentions a recipe she describes as clay and ash chicken.[1]

Many animals involuntarily ingest clay with their food in pastures without incurring fatal consequences. However, it is noted that some of them seek a specific type of clay and are able to select it to cope with intestinal disorders or bear famine periods. This additive has a digestive or functional role, empirically recognized, which is becoming better known scientifically[2] (Reichardt 2008). The animal feed industry largely uses various clays (kaolinite, smectite, sepiolite, palygorskite) with an apparent appreciable success to improve productivity and breeding quality.

It is dramatic to note that as we are writing this chapter, there still remain many geographic regions on our planet where famine subsists. We do not want to pass judgment here, but only point out that, for example, in Haiti (Fig. 8.2) as in too many other regions of the world, unfortunately subjected to serious conflicts or climatic disasters, the population is forced to consume "soil cakes," in fact made of clay, in order to alleviate their hunger.[3]

8.2.2 Application to Functional or Temporary Intestinal Disorders

It is necessary to present the environment in which clay acts. It involves the entire digestive system: mouth, esophagus, stomach, digestive tract (the intestinal lumen and the external wall of the mucous membranes of the digestive system), and down to the anus (Fig. 8.8). Very varied situations exist, so we consider those that are general throughout this system.

---

[1] Extract of the recipe described by Colette (translation):

> … It seems barbarian. It points out that Chinese chicken, sealed in the lacquer, except that the chicken with ash requires it to be eliminated, wrapped in smooth clay, the clay of sculptors. One should only gut it out carefully, pepper it and salt it internally. Its grease, captive is sufficient. The clay ball and its coated chicken undergo a rather long cremation within thick ash, all around surrounded by embers poked and renewed. The soft clay, after three quarters of an hour, has become a terra cotta egg. Break it: all the feathers, part of the skin, remain attached to the shards, and the wild perfection of a tender chicken inclines you towards rather brutal and prehistoric greediness…

[2] Parrots from southeast Peru are accustomed to consuming earth charged with kaolinite. Gilardi et al. (1999) shows that it is a question for them of fighting against the toxicity of seeds consumed before their maturity. The alkaloids contained in the seeds (protection that enables trees to reproduce by prohibiting premature seed consumption) are fixed by the kaolinite particles (mechanism of chelation that protects the animals). The kaolinite covers their intestinal wall, thus adding to protection. Moreover, rock salt can be consumed on this occasion.

[3] Emissions: TV news télé2 of France 2 of April 2008 and the emission "Thalassa," December 12, 2008, inter alia.

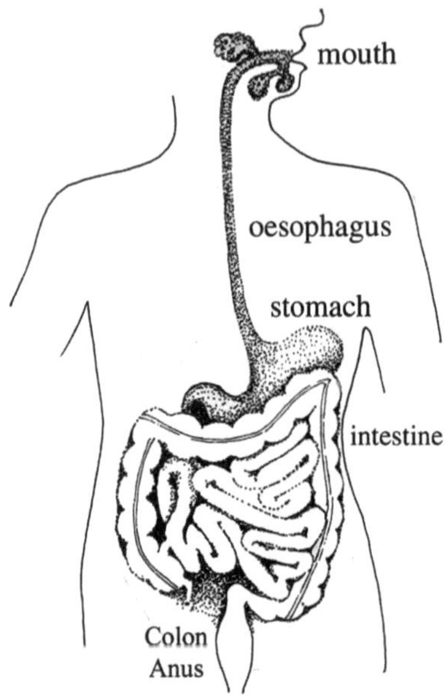

Fig. 8.8 Simplified diagram of the digestive system

Mucous membranes are made of epithelial cells covered with a continuous layer of viscous mucus secreted by some of these cells. Mucus coats the stomach surface with a 1–1.5 mm thick film that protects the mucous membrane against hydrochloric acid and prevents cellular autodigestion by the digestive enzymes. It is described as a gel whose viscosity characteristics are particularly important for its effective function. These viscosity parameters and composition are fragile, taking into account varied and aggressive environmental conditions. The mucus forms a protective barrier that is unceasingly renewed by secretion but constantly eliminated by the matter that circulates in a natural way in the digestive light. The mucus barrier has a physical effect against the chemical or physical aggressions of the substances and molecules in transit; it has a buffer effect that adapts the pH and protects the mucous membranes from direct contact with intestinal flora. This sequence is sometimes disturbed by dysfunctions or disorders of certain bodies but also by the ingestion of badly assimilated or harmful products. The protection of the mucous membrane is then weakened which causes functional disordered states on various levels of the digestive chain.

Various methods of safeguarding or protecting the mucous membranes' state are possible: clay ingestion is one. Clay adsorption properties are generally taken into account; however, in fact, it is the partial coating of the mucous membrane that occurs.

The mixture of clay and mucus spread at the surface of the intestinal villi forms a reinforced mechanical barrier that protects the mucous membrane against medium aggressiveness; the intestinal mucous membrane can thus better fulfill its role and ensure a correct transit (Fig. 8.9). Because of their high surface reactivity clay particles can then react with polar molecules or free radicals, and be particularly sensitive to the polarized nature of the membrane epithelial cells.

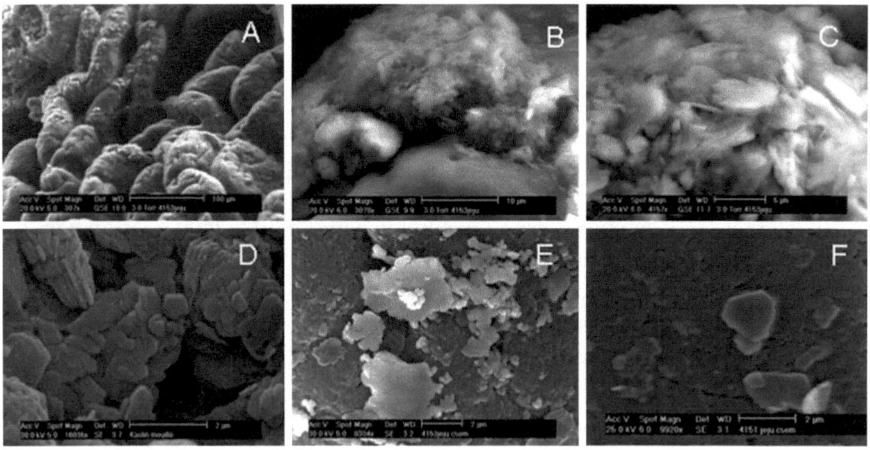

Fig. 8.9 Intestinal wall of a rat having ingested kaolinite; observations under the scanning electron microscope under controlled vacuum pressure. *Hydrated samples under weak vacuum*: (**a**) villi tops; (**b**) clay coat at the level of the mucous membrane; (**c**) detail of the clay particle stacking. *Dehydrated samples under high vacuum pressure*: (**d**) platy particles of the reference clay observed before ingestion; (**e**, **f**) tiny clay plates after ingestion

Independently of its action on the mucosal layer, clay is also mixed with the mass of matter in transit with which it could be introduced during food absorption. In such a complex medium, it has many possibilities of action, for example, the fixing of harmful molecules such as toxins which are always present in the digestive environment and come from the activity of the inevitable microorganisms. The latters and some viruses can be adsorbed by clay that in this way reduces their action on the mucous membranes and their activity in general. The action of clay also appears in the treatment (absorption or elimination) of superabundant stomachic or intestinal gases resulting from the processes of digestion.

All these properties are arguments in favor of the use of clay in the treatment of gastrointestinal dysfunctions. The return to normal functions that often results is due to the mucus properties (quantity and viscosity). A relation between the time of retention of a given material (molecule, gas, microorganism, etc.) on the clay and the speed of intestinal transit time must be well established to understand and control its action. On the whole the action of clay is clearly recognized and can even make it possible to control some severe ulcerous ailments.

However, we have seen that the word "clay" is too generic and covers families, species, and varieties whose properties are very diversified. They cannot be managed without knowing their nature and properties precisely. Another important point is that of the associated minerals, always very finely divided, difficult, even impossible, to separate; they can have a considerable active role. The cells constituting the external brush surface of the intestinal membrane are sufficiently sensitive to be reactive to these materials characterized mainly by their small dimensions that classify them in the colloidal field and are therefore very active from the physicochemical point of view.

8.3 Clay Applications in the Presence of Other Active Molecules

Using clay simultaneously with other molecules, special attention must be paid to avoid unexpected contrary effects, such as "certain greasy substances" or "drugs," for example.

In internal use it can prove useful to employ paraffin oil to treat constipation, for instance. In this case clay and paraffin oil should never be simultaneously employed because the adsorption of oil on clay leads to an extremely harmful hardening. For paraffin oil ingestion, it is necessary to observe a period of several days before beginning a treatment containing clay in order to make sure of the total elimination of oil and vice versa.

The adsorption of the active molecules of one drug on the surface of clay can be sufficiently strong to stop or decrease its action. Treatments containing clay and those based on pharmaceutical molecules must be used separately with an interval of a few days.

References

Abrahams PW (2005) Geophagy and the involuntary ingestion of soils. In: Selinu O, Alloway B, Centeno AA, Finkelman RB, Fuge R, Lindh U, Smedley P (eds) Essentials of medical geology. Elsevier Academic Press, London, pp 435–458

Aguzzi C, Cerezo P, Viseras C, Caramella C (2007) Use of clays as drug delivery system: possibilities and limitations. Appl Clay Sci 36:22–36

Araújo ARTS, Paiva T, Ribeiro MP, Coutinho P (2015) Innovation in thermalism: an example in Beira interior region of Portugal, Chapter 11. In: Peris-Ortiz M, Álvarez-Garcia J (eds) Health and wellness tourism: emergence of a new market segment. Springer, Switzerland, pp 165–180

Armijo F (1991) Propiedades térmicas de los peloides. Bol Soc Esp Hidrol Med 6(3):151–158

Armijo F (2015) Fases sólidas de los peloides: propiedades térmicas y mecânicas. Balnea 10:143–162

Arribas M, Meijide R, Mourelle ML (2010a) Evolución de la psoriasis tratada con peloides y água mineromedicinal de La Toja. In: Maraver F, Carretero MI (eds) Libro de Resúmenes del II Congreso Iberoamericano de Peloides. C.E.R.S-A, Madrid, p 75

Arribas M, Meijide R, Mourelle ML, Gómez CP, Legido JL (2010b) Projecto de innovación para el desarrollo de un peloide termal para el balneario isla de la Toja. In: Maraver F, Carretero MI (eds) Libro de Resúmenes del II Congreso Iberoamericano de Peloides. C.E.R.S-A, Madrid, p 76

Arribas M, Mejide R, Mourelle ML (2012) Long-term effect of psoriasis treatment with mineral water and peloids of La Tora. Balnea 6:289–290

Artymuk NV, Kira EF, Kondratieva TA (2010) Intravaginal gel prepared from Dead Sea peloid for treating luteal-phase defect. Int J Gynaecol Obstet 108(1):72–73

Baby AR, Zague V, Maciel CPM, Salgado-Santos KY, Arêas EPG, Kaneko TM, Consiglieri VO, Velasco MVR (2004) Development of cosmetic mask formulations. Braz J Pharm Sci 40(1):159–161

Bagnato G (2004) Clinical improvement and serum amino acid levels after mud-bath therapy. Int J Clin Pharmacol Res 24:39–47

Banenzoue C, Signing P, Mbey JA, Njopwouo D (2014) Antacid power and their enhancements in some edible clays consumed by geophagia in Cameron. J Chem Pharm Res 6(10):668–676
Barros M, Santos D, Pena Ferreira MR, Silva JBP, Amaral MH, Sousa Lobo JM, Gomes JHCA, Gomes CSF (2010) Estudos de pré-formulação de um pelóide anticelulitico com argila esmectítica da ilha do Porto Santo. In: Maraver F, Carretero MI (eds) Libro de Resúmenes del II Congreso Iberoamericano de Peloides. C.E.R.S-A, Madrid, pp 77–78
Baschini MT, Pettinari GR, Vallés JM, Aguzzi C, Cerezo P, López-Galindo A, Setti M, Viseras C (2010) Suitability of natural sulphur-rich muds from Copahue (Argentina) for use as semisolid health care products. Appl Clay Sci 49:205–212
Baschini MT, Piovano E, López-Galindo A, Dietrich D, Setti M (2014) Composición y propriedades de fangos (peloides), águas y sales procedentes de lagunas y lagos salinos usados com fines terapêuticos y cosméticos. In: Torres HA (coordenador) Peloteraia: Aplicaciones médicas y cosméticas de fangos termales. Fundación Bibilis, pp 145–154
Bech J (1987) Les terres médicinales. Discurs per Reial Académia de Farmácia de Barcelona. Reial Académia de Farmácia de Barcelona-CIRIT (Generalitat de Catalunya), Barcelona
Beer AM, Junginger HE, Lukanov J, Sagorchev P (2003) Evaluation of the permeation of peat substances through human skin in vitro. Int J Pharm 253:169–175
Beer AM, Fetaj S, Lange U (2013) Peloid therapy: an overview of the empirical status and evidence of mud therapy. Z Rheumatol 72:581–589
Bellometti S, Cecchettin M, Galzigna L (1997) Mud pack therapy in osteoarthrosis. Changes in serum levels of chondrocyte markers. Clin Chim Acta 268(1–2):101–106
Bellometti S, Poletto M, Gregotti C, Richelmi P, Bertè F (2000) Mud bath therapy influences nitric oxide, myeloperoxidase and glutathione peroxidase serum levels in arthritic patients. Int J Clin Pharmacol Res 20(3–4):69–80
Bellometti S, Galzigna L, Richelmi P, Gregotti C, Bertè F (2002) Both serum receptors of tumor necrosis factor are influenced by mud pack treatment in osteoarthrotic patients. Int J Tissue React 24:57–64
Bellometti S, Richelmi P, Tassoni T, Bertè F (2005) Production of matrix metalloproteinases and their inhibitors in osteoarthritic patients undergoing mud bath therapy. Int J Clin Pharmacol Res 25(2):77–94
Bellometti S, Gallotti C, Pacileo G, Rota A, Tenconi MT (2007) Evaluation of outcomes in spa-treated osteoarthrosic patients. J Prev Med Hyg 48(1):1–4
Bender T, Karagülle Z, Bálint G, Gutenbrunner C, Bálint P, Sukenik S (2005) Hydrotherapy, balneotherapy, and spa treatment in pain management. Rheumatol Int 25:220–224
Benedetti S, Pagliarani S, Benvenuti F, Marini D, Galli T, Oliva F, Lazzari P, Canestrari F (2007) Antioxidative effects of sulphurous water from Macerata Feltria Thermal Resort in patients with osteoarthritis. Prog Nutr 9(1):46–52
Bolzinger MA, Briançon S, Pelletier J, Chevalier Y (2012) Penetration of drugs through skin, a complex rate-controlling membrane. Curr Opin Colloid Interface Sci 17(3):156–165
Burguera EF, Vela-Anero A, Magalhães J (2012) Effect of hydrogen sulfide sources on inflammation and catabolic markers on interleukin 1β-stimulated human articular chondrocytes. Osteoarthritis Cartilage 22:1026–1035
Cantarini L, Leo G, Giannitti C, Cevenini G, Barberini P, Fioravanti A (2007) Therapeutic effect of spa therapy and short wave therapy in knee osteoarthritis: a randomized, single blind, controlled trial. Rheumatol Int 27:523–529
Cara S, Carcangiu G, Padalino G, Palomba M, Tamanini M (2000) The bentonites in pelotherapy: thermal properties of clay pastes from Sardinia (Italy). Appl Clay Sci 16:125–132
Carbajo JM, Maraver F (2015) Hydrogen sulfide and health. New insights. Balnea 10:93–105
Cardoso-Gomes J, Gomes CSF (2015) Mud used for therapeutic and skin care purposes at the beach of Porto de Mós, Algarve, Portugal. Balnea 10:355–356
Carretero MI (2002) Clay minerals and their beneficial effects upon human health: a review. Appl Clay Sci 21(3–4):155–163

Carretero MI (2013) Metodologia aplicada a la determinación del peloid óptimo. Livro de Actas do III Congresso Iberoamericano de Peloides. In: Nunes J, Gomes C, Silva J (eds). Ponta Delgada, São Miguel, Açores, pp 19–24

Carretero I, Pozo M (2007) Mineralogia Applicada: Saud y Medio Ambiante. Thomson Editors, Madrid, p 406

Carretero MI, Gomes CSF, Tateo F (2006) Clays and human health. In: Bergaya F, Theng BKG, Lagaly G (eds) Handbook of clay science, vol 1, Developments in clay science. Elsevier, Amsterdam, pp 717–741

Carretero MI, Pozo M, Martin-Rubi JA, Pozo E, Maraver F (2010) Mobility of elements in interaction between artificial sweat and peloids used in Spanish spa. Appl Clay Sci 48(3):506–515

Carretero MI, Gomes CSF, Tateo F (2013) Clays and human health. In: Bergaya F, Lagaly G (eds) Handbook of clay science, vol 1, 2nd edn, Developments in clay science. Amsterdam, Elsevier, pp 711–764

Carretero MI, Pozo M, Legido JL, Fernandez-González MV, Delgado R, Gómez I, Armijo F, Maraver F (2014) Assessment of three Spanish clays for their use in pelotherapy. Appl Clay Sci 99:131–143

Casas LM, Pozo M, Gómez CP, Pozo E, Bessieres D, Plantier F, Legido JL (2013) Thermal behaviour of mixtures of bentonitic clay with different salinity waters. Appl Clay Sci 72:18–25

Ceccarelli F, Perricone C, Alessandri C, Modesti M, Lagnocco A, Croia C, Di Franco M, Valesini G (2010) Exploratory data analysis on the effects of non-pharmacological treatment for knee osteoarthritis. Clin Exp Rheumatol 28:250–253

Codish S, Abu-Shakra M, Flusser D, Friger M, Sukenik S (2005) Mud compress therapy for the hands of patients with rheumatoid arthritis. Rheumatol Int 25:49–54

Costantino M (2006) Sulphur mud-bath treatment in osteoarthrosis: therapeutic activity and efficiency on the quality of life. Clin Ter 157:525–529

Cozzi F, Carrara M, Sfriso P, Todesco S, Cima L (2004) Anti-inflammatory effect of mud-bath applications on adjuvant arthritis in rats. Clin Exp Rheumatol 22:763–766

Curri SB, Bombardelli E, Grossi F (1997) Observations on organic components of thermal mud: morphohistochemical and biochemical studies on lipid components of mud of the Terme dei Papi (Laghetto del Bagnaccio, Viterbo). Chemical bases of the interpretation of biological and therapeutic actions of thermal mud. Clin Ter 148:637–654

De Michele D, Giacomino M, Untura-Filho M, Beldarrain A (2006) Efectos sistémicos de los fangos minerales. Revisión de la literatura de los últimos 10 años [Systemics effects of minerals muds. A ten years literature review]. An Hidrol Méd 1:135–142

De Vos P (2010) European material medica in historical texts: longevity of a tradition and implications for future use. J Ethnopharmacol 176:10–17

Denaverre M (1975) Face masks (in Denaverre M), the chemistry and manufacture of cosmetics, vol 3, 2nd edn. Continental Press, Orlando

Espejo-Antúnez L, Cardero-Durán MA, Garrido-Ardila EM, Torres-Piles S, Caro-Puértolas B (2013) Clinical effectiveness of mud pack therapy in knee osteoarthritis. Rheumatology (Oxford) 52:659–668

Evcik D, Kavuncu V, Yeter A, Yigit I (2007) The efficacy of balneotherapy and mud-pack therapy in patients with knee osteoarthritis. Joint Bone Spine 74:60–65

Faílde RM, Mosquera LM (2006) Afecciones dermatológicas y cosmética dermotermal. Técnicas y Tecnologias en Hidrologia Médica y Hidroterapia 175–179

Fernández-Gonzalez MV, Martin-Garcia JM, Delgado G, Párraga J, Carretero MI, Delgado R (2013) Una aproximación al estabelecimento de los tempos óptimos de maturación. Livro de Actas do III Congresso Iberoamericano de Peloides. In: Nunes J, Gomes C, Silva J (eds). Ponta Delgada, São Miguel, Açores, pp 127–131

Fernández-Lao C, Cantarero I, Garcia JF, Arroyo M (2012) Termoterapia. In: Albornoz M, Meroño J (eds) Procedimientos generales de fisioterapia. Elsevier, Barcelona, pp 53–65

Ferrand T, Yvon J (2000) Thermal properties of clay pastes for pelotherapy. Appl Clay Sci 6:751–760

Ferrell RE (2008) Medicinal clay and spiritual healing. Clay Clay Miner 56:751–760

Fioravanti A, Chelesschi S (2015) Mechanisms of action of balneotherapy in rheumatic diseases. Balnea 10:43–56

Fioravanti A, Lacoponi F, Bellisai B, Cantarini L, Galeazzi M (2010) Short-term and long-term effects of spa therapy in knee osteoarthritis. Am J Phys Med Rehabil 89:125–132

Fioravanti A, Cantarini L, Guidelli GM, Galeazzi M (2011) Mechanisms of action of spa therapies in rheumatic diseases: what scientific evidence is there? Rheumatol Int 31(1):1–8

Fioravanti A, Lamboglia A, Pascarelli NA (2013) Thermal water of Vetriolo, Trentino, inhibits the negative effect of interleukin 1β on nitric oxide production and apoptosis in human osteoarthritic chondrocytes. J Biol Regul Homeost Agents 27:891–902

Fioravanti A, Tenti S, Gianitti C, Fortunati NA, Galeazzi M (2014) Short and long-term effects of mud-bath treatment on hand osteoarthritis: a randomized clinical trial. Int J Biometeorol 58(1):79–86

Flusser D, Abu-Shakra M, Friger M, Codish S, Sukenik S (2002) Therapy with mud compresses for knee osteoarthritis: comparison of natural mud preparations with mineral-depleted mud. J Clin Rheumatol 8(4):197–203

Fouquey M (2014) Boues et Guide des Bonnes Pratiques Thermales. AFth-Bulletin d'Information 25:12–16

Fraioli A, Serio A, Mennuni G, Ceccarelli F, Petraccia L, Fontana M, Grassi M, Valesini G (2010) A study on the efficacy of treatment with mud packs and baths with Sillene mineral water (Chianciano Spa Italy) in patients suffering from knee osteoarthritis. Rheumatol Int. doi:10.1007/s00296-010-1475-5

Francois G, Micollier A, Rouvie I (2005) Les Boues Thermales. Atelier Santé Environmental, ENSP (École Nationale de la Santé Publique), Rennes, 29 pp

Galzigna L, Moretto C, Lalli A (1996) Physical and biochemical changes of thermal mud after maturation. Biomed Pharmacother 50(6–7):306–308

Galzigna L, Bettero A, Bellometti S (1999a) La maturation de la boue thermale et sa mesure. Première partie. Press Therm Clim 136(1):23–26

Galzigna L, Bettero A, Bellometti S (1999b) La maturation de la boue thermale et sa mesure. Deuxième partie. Press Therm Clim 136(1):27–30

Garcia PT (2014) Peloterapia en Cosmética y Medicina Estética. In: Torres AH (coordinador) Peloteraia: Aplicaciones médicas y cosméticas de fangos termales. Fundación Bílbilis, pp 185–207

Gilardi JD et al (1999) Biochemical functions of geophagy in parrots: detoxification of dietary toxins and cytoprotective effects. J Chem Ecol 25:897–922

Godin B, Touitou E (2007) Transdermal skin delivery: predictions for humans from in vivo, ex vivo and animal models. Adv Drug Deliv Rev 59:1–10

Gomes CSF (2002) Argilas: Aplicações na Indústria. O Liberal, Câmara de Lobos, R.A. da Madeira, 377 pp

Gomes CSF (2015) In pelotherapy what is more important, the peloid solid phase or the peloid liquid phase? Balnea 10:125–142

Gomes CSF, Rautureau M (2013) The role of clay and clay minerals in peloid properties. Livro de Actas do III Congresso Iberoamericano de Peloids. Ponta Delgada, São Miguel, Açores, pp 86–93

Gomes CSF, Silva JBP (2007) Minerals and clay minerals in medical geology. Appl Clay Sci 36:4–21

Gomes CSF, Hernandez R, Sequeira MC (2008) Characterization of clays used for medicinal purposes in the Archipelago of Cape Verde. Geochim Bras 22(3):315–331

Gomes CSF, Carretero MI, Pozo M, Maraver F, Cantista P, Armijo F, Legido JL, Teixeira F, Rautureau M, Delgado R (2013a) Peloids and pelotherapy: historical evolution, classification and glossary. Appl Clay Sci 75–76:28–38

Gomes CSF, Rautureau M (2013) The role of clay and clay minerals in peloid properties. Livro de Actas do III Congresso Iberoamericano de Peloids, Nunes J, Gomes C, Silva J. (editores), Ponta Delgada, São Miguel, Açores, 86-93

Gomes A, Simões M, Cavaleiro J, Gomes J, Gomes C, Silva J, Santos D, Pontes A (2013b) Study of the organic constituents of the volcanic mud from the "blue pool" of the Balneário da Coroa, São Miguel Island, Azores, Portugal. In: Nunes J, Gomes C, Silva J (eds) Livro de Actas do III Congresso Iberoamericano de Peloids. Ponta Delgada, São Miguel, Açores, pp 332–339

Gomes CSF, Silva JBP, Gomes JHC (2015) Natural peloids versus designed and engineered peloids. Bol Soc Esp Hidrol Méd 30(1):15–36
Gupta S, Bansal R, Gupta S, Jindal N, Jindal A (2013) Nanocarriers and nanoparticles for skin care and dermatological treatments. Indian Dermatol Online J 4(4):267
Hattori I (1963) Pelotherapy. In: Sidney Licht MD (ed) Medical hydrology. Waverly Press, Baltimore, pp 273–290
Horno MAP (2014) Historia de la Peloterapia. In: Torres AH (coordinador) Peloterapia: Aplicaciones médicas y cosméticas de fangos termales. Fundación Bílbilis, pp 47–53
Khanna N, Gupt SD (2002) Rejuvenating facial massage—a bane or boom. Int J Dermatol 41:407–410
Kikouama OJR, Baldé L (2010) From edible clay to clay-containing formulation for optimization of the oral delivery of some trace elements: a review. Int J Food Sci Nutr 61(8):1–21
Kikouama OJR, Le Cornec F, Bouttier S, Launay A, Baldé L, Yagoubi N (2009a) Evaluation of trace elements released by edible clays in physicochemically simulated physiological media. Int J Food Sci Nutr 60(2):130–142
Kikouama OJR, Konan KL, Katty A, Bonnet JP, Baldé L, Yagoubi N (2009b) Physicochemical characterization of edible clays and release of trace elements. Appl Clay Sci 43(1):135–141
Kim JH, Lee J, Lee HB, Shin JH, Kim EK (2010) Water-retentive and anti-inflammatory properties of organic and inorganic substances from Korean sea mud. Nat Prod Commun 5(3):395–398
Legido JL, Mourelle ML (eds) (2008) Investigaciones en el ámbito Iberoamericano sobre Peloides termales. Universidad de Vigo, 308 pp
Lim DG, Jeong WW, Kim NA, Lim JY, Lee SH, Shim WS, Kang NG, Jeong SH (2014) Effect of the glyceryl monooleate-based lyotropic phases on skin permeation using in vitro diffusion and skin imaging. Asian J Pharm Sci 9:324–329
Liu H, Zeng C, Gao SG, Yang T, Luo W, Li YS (2013) The effect of mud therapy on pain relief in patients with knee osteoarthritis: a meta-analysis of randomized controlled trials. J Int Med Res 41:1418–1425
Lopez-Galindo A, Viseras C (2004) Pharmaceutical and cosmetic application of clays. In: Wypych F, Satyanarayana KG (eds) Clay surfaces: fundamentals and applications. Elsevier, Amsterdam, pp 267–289
Lopez-Galindo A, Viseras C, Cerezo P (2007) Compositional, technical and safety specifications of clays to be used as pharmaceutical and cosmetic products. Appl Clay Sci 36:51–63
Mahboob N, Sousan K, Shirzad A, Amir G, Mohammad V, Reza M, Mansour VA, Hadi V (2009) The efficacy of a topical gel prepared using Lake Urmia mud in patients with knee osteoarthritis. J Altern Complement Med 15:1239–1242
Maraver F (2006) Antecedentes históricos de la Peloterapia. An Hidrol Méd 1:17–42
Maraver F (2008) El agua mineromedicinal: Una herramienta curativa milenaria. Tribunatermal 11:54–57
Maraver F (2013) Mechanisms of action of pelotherapy: state of the art. In: Nunes J, Gomes C, Silva J (eds) Livro de Actas do III Congresso Iberoamericano de Peloides. Ponta Delgada, São Miguel, Açores, pp 9–18
Maraver F, Fernandez-Torán MA, Corvillo I, Morer C, Váquez I, Aguillera L, Armijo F (2015) Peloterapia: Una Revisión. Med Nat 9(1):38–46
Matz H, Orion E, Wolf R (2003) Balneotherapy in dermatology. Dermatol Ther 16(2):132–140
Meijide R, Salgado T, Lianes A, Legido JL, Mourelle ML, Gómez C (2010) Evaluación de los câmbios en la piel trás la aplicación de peloides mediante métodos de bioengenharia cutânea. In: Maraver F, Carretero MI (eds) Libro de Resúmenes del II Congreso Iberoamericano de Peloides. C.E.R.S.-A, Madrid, pp 48–49
Meijide R, Mourelle ML, Vela-Anero A, López EM, Burguera EF, Pérez CG (2014) Aplicación a pacientes: Peloterapia en patologias dermatológicas. In: Peloterapia: Aplicaciones médicas y cosméticas de fangos termales. Fundación Bílbilis, Coordinator: Torres AH, pp 169–183
Meijide R, Burguera EF, Vela-Anero A (2015) Peloterapia y Artrosis. Balnea 10:289–300
Naik A, Pechtold L, Potts R, Guy R (1995) Mechanism of oleic acid-induced skin penetration enhancement in vivo in humans. J Control Release 37:299–306

Nissenbaum A, Rullköetter J, Yechieli Y (2002) Are the curative properties of 'black mud' from the Dead Sea due to the presence of bitumen (asphalt) or other types of organic matter? Environ Geochem Health 24(4):327–335

Novelli G (1996) Applicazion medicali e Igieniche delle bentoniti. Atti Conv. "Argille Curative". In: Veniale F (ed) Gruppo Italiano AIPEA, Salice Terme (PV), Tipografia Trabella, Milano

Novelli G (1998) Applicazioni Cosmetiche e Medicaliu delle argille smectiche. Cosmetic News 122:350–357

Novelli G (2000) Bentonite: a clay over the centuries. Incontri Scentifici, V Corso di Formazione "Metodi di Analisi di Materiali Argillosi". Gruppo Itaiano AIPEA, pp 263–304

Odabasi E, Gul H, Macit E (2007) Lipophilic components of different therapeutic mud species. J Altern Complement Med 13:1115–1118

Odabasi E, Turan M, Erdem H, Pay S, Guleç M, Karagulle MZ (2009) The effect of mud pack treatment in knee osteoarthritis. Turk J Rheumatol 24:72–76

Otto CC, Haydel SE (2013) Exchangeable ions are responsible for the in vitro antibacterial properties of natural clay mixtures. PLOS ONE/ www.plosone, 2013;8,(5),1–9

Pandey A, Mittal A, Chauhan N, Alam S (2014) Role of surfactants as penetration enhancer in transdermal drug delivery. J Mol Pharm Org Process Res 2:2

Panico V, Imperato R (2009) The psoriasis: a therapeutic alternative with sulphurous water of Terme Capasso. J Water Welness 1(1):39–50

Pastor JM (1998) Termoterapia superficial. In: Martínez M, Pastor JM, Sendra F (eds) Manual de Medicina Fisica. Harcourt Brace, Madrid, pp 91–104

Pena-Ferreira MR, Santos D, Silva JBP, Amaral MH, Sousa Lobo JM, Gomes JHCA, Gomes CSF (2010a) Aplicação de argilas esmectíticas da ilha do Porto Santo em máscaras faciais. In: Maraver F, Carretero MI (eds) Libro de resúmenes del II Congreso Iberoamericano de Peloides. C.E.R.S-A, Madrid, pp 30–31

Pena-Ferreira MR, Santos D, Silva JBP, Amaral MH, Sousa Lobo JM, Gomes JHCA, Gomes CSF (2010b) Desenvolvimento de formulações contendo argila esmectítica e areia carbonatada biogénica da ilha do Porto Santo para aplicação em máscaras de limpeza e branqueadoras. In: Maraver F, Carretero MI (eds) Libro de resúmenes del II Congreso Iberoamericano de Peloides. C.E.R.S-A, Madrid, pp 99–100

Pena-Ferreira MR, Santos D, Silva JBP, Amaral MH, Sousa-Lobo JM, Gomes JH, Gomes CSF (2011) Aplicações de argilas esmectíticas da ilha do Porto Santo em máscaras faciais. An Hidrol Méd 4:67–79

Porlezza C (1965) Considerazione sui fanghi terapeutici (peloidi). Thermae II(2–3):6–57

Pozo M (2015) Importancia de la composición y propriedades físicas de las arcillas en la preparación de peloides. Balnea 10:187–206

Pozo M, Carretero MI, Pozo E, Marin Rubi JA, Maraver F (2010) Caracterización mineralógica y química de peloides españoles y argentinos: Evaluación de elementos traza potencialmente tóxicos. En: Maraver F, Carretero MI (eds) Libro de Resúmenes del II Congreso Iberoamericano de Peloides, Lanjarón, pp 37–38

Pozo M, Carretero MI, Maraver F, Pozo E, Gómez I, Armijo F (2013) Composition and physical-physicochemical properties of peloids used in Spanish spas: a comparative study. Appl Clay Sci 83–84:270–279

Rambaud A (1989) Les boues thermales. In: Hérisson C (ed) Crénotherapie et Réadaptation. Masson, Paris, pp 9–19

Rambaud A, Rambaud J, Berger G, Pauvert B (1986) Mesure et étude du comportement thermique des boues thermales. J Fr Hydrol 17:392–302

Rautureau M, Liewig N, Gomes CSF, Katouzian-Safadi M (2010) Argiles et Santé: Propriétés et Thérapies. Editions Médicales Internationales, Lavoisier, 184 pp. ISBN 978-2-7430-1202-1

Reichardt F (2008) Ingestion d'argile spontanée chez le rat: rôle dans la physiologie intestinale. Thèse, Université Louis Pasteur, Strasbourg

Reinbacher R (2003) Healing earths: the third leg of medicine. 1st Books Library, Bloomington, IN, 244 pp

Rondanelli M, Opizzi A, Perna S, Faliva MA, Buonocore D, Pezzoni G, Michelotti A, Marchetti R, Marzatico F (2012) Significant two-weeks clinical efficacy of an association between Massaciuccoli peat and sodium chloride water of Undulna Thermae measured on gynoid lipodystrophy in a group of overweight female. Ann Ig 24(5):369–378
Roques CF (2004) Mud therapy and health. Proceedings of the 3rd symposium on thermal muds in Europe, Dax, pp 75–77
Roques CF (2015) Mud therapy: data for clinical evidence. Balnea 10:57–62
San Martin Bacaicoa J (1994) Peloides en general: Características físicas, efectos biológicos e indicaciones terapéuticas. In: Armijo M, San Martin J (eds) Curas balneárias y climáticas, Talasoterapia y Helioterapia. Complutense, Madrid, pp 315–331
Sanchez CJ, Parras J, Carretero I (2002) The effect of maturation on the mineralogical and physicochemical of illitic-smectitic clays for pellotherapy. Clay Miner 37:457–464
Sarsan A, Akkaya N, Ozgen M, Yildiz N, Atalay NS, Ardic F (2012) Comparing the efficacy of mature mud pack and hot pack treatments for knee osteoarthritis. J Back Musculoskelet Rehabil 25(3):193–199
Schaefer H, Redelmeier TE (1996) Skin barrier: principles of percutaneous absorption. In: Karger S (ed) Zurich, Switzerland
Schnetz E, Fartasch M (2002) Microanalysis for the evaluation of penetration through the human skin barrier: a promising tool for future research? Eur J Pharm Sci 12:165–174
Serofilli A (1994) La Fango—Balneoterapia. PICCIN, Padova, 130 pp
Silva JBP, Santos D, Amaral MH, Ramos P, Franco A, Gomes CSF, Lobo JM (2015) Casos de estudo de tratamento de patologia mamária com peloide terapêutico. Balnea 10:120–122
Steinstand S (2012) Transdermal delivery of water soluble molecules into human skin. MSc Thesis, Norwegian University of Science and Technology (NTNU), Trondheim, 131 pp
Suárez M, González P, Dominguez R, Bravo A, Melián C, Perez M (2011) Identification of organic compounds in San Diego de los Baños peloid (Pinar del Rio, Cuba). J Altern Complement Med 17(2):155–165
Suresh R, Borkar SN, Sawant VA, Shende VS, Dimble SK (2010) Nanoclay drug delivery systems. Int J Pharm Sci Nanotechnol 3(2):901–905
Tanojo H, Boelsma E, Junginger H, Ponec M, Boddé H (1999) In vivo human skin permeability enhancement by oleic acid: a laser Doppler velocimetry study. J Control Release 58:97–104
Tateo F, Summa V (2007) Element mobility in clays for healing use. Appl Clay Sci 36:64–76
Tateo F, Ravaglioli A, Andreoli C, Bonina F, Coiro V, Degetto S, Giaretta A, Menconi Orsini A, Puglia C, Summa V (2009) The in vitro percutaneous migration of chemical elements from a thermal mud for healing use. Appl Clay Sci 44:83–94
Tefner IK (2014) Effects of balneotherapy on musculoskeletal disorders with chronic pain. Ph.D. Thesis, Univ. Szeged, Hungary, 80 pp
Teixeira F, Maraver F, Crespo PV, Campos A (1996) Estudo microanalítico da materia orgánica de águas sulfúreas portuguesas e espanholas. Pub Inst Clim Hidrol da Univ de Coimbra 34:1–5
Tenti S, Cheleschi S, Galeazzi M, Fioravanti A (2014) Spa therapy: can be a valid option for treating knee osteoarthritis? Int J Biometeorol. doi:10.1007/s00484-014-0913-6.5
Tolomio C, Ceschi-Berrini C, Moschin E, Galzigna L (1999) Colonization by diatoms and antirheumatic activity of thermal mud. Cell Biochem Funct 17(1):29–33
Torrella F (2006) La sulfuraria de Baños de Montemayor (Cáceres): características morfológicas y funcionales de la comunidad microbiana constituyente. An Hidrol Med 1:61–78
Tricás JM, Fortún M, Jiménez S, Estébanes E (2014) Fisioterapia: fundamentación fisioterápica de la utilización de peloides. In: Hernandez A (coord.). Peloterapia: aplicaciones médicas y cosméticas de fangos termales, Madrid, Fundación Bilbilis, pp 251–260
Tserenpil S, Dolman G, Voronkov MG (2010) Organic matters in healing muds from Mongolia. Appl Clay Sci 49(1–2):55–63
Ubogui J, Rodríguez-Lupo L, Ficoseco H, Sevinsky L, Kien K, Stengel F (1991) Terapéutica no convencional de la Psoriasis en las termas de Copahue (Neuquén Argentina). Experiencia preliminar. Arch Arg Dermatol 41:25–39

Ubogui J, Stengel FM, Kien MC, Sevinsky L, Rodríguez Lupo L (1998) Thermalism in Argentina: alternative or complementary dermatologic therapy. Arch Dermatol 134(11):1411–1412
Vaht M, Birkenfeldt R, Ubner M (2008) An evaluation of the effect of differing lengths of spa therapy upon patients with osteoarthritis (OA). Complement Ther Clin Pract 14:60–64
Varga C (2010) Problems with classification of spa waters used in balneology. Health 2(11):1260–1263. doi:10.4236/health.2010.211187
Veniale F (ed) (1996) Atti di Convegno Argille Curative. Associazione Medica Italiana di Idroclimatologia, Talassologia e Terapia Física, Gruppo Italiano AIPEA. Salice Terme /PV, 145 pp
Veniale F (1998) Applicazioni e utilizzazioni medico-sanitarie di materiali argillosi (naturali e modificati). Corso di Specializzazione, Gruppo Italiano AIPEA, pp 1–40
Veniale F (ed) (1999) Argille curative per fanghi peloidi termali e per trattamienti dermatologici e cosmetici. Symp.Gruppo Ital. AIPEA, Montecatini Terme/PT, Miner. Petrogr. Acta, v. XLII, 261–335
Veniale F, Barberis E, Carcangiu G, Morandi N, Setti M, Tamanini M, Tessier D (2004) Formulation of muds for pelotherapy: effects of "maturation" by different mineral waters. Appl Clay Sci 25:135–148
Viseras C, López-Galindo A (1999) Pharmaceutical applications of some Spanish clays (sepiolite, palygorskite, bentonite): some pre-formulation studies. Appl Clay Sci 14:69–82
Viseras C, Cultrone G, Cerezo P, Aguzzi C, Baschini M, Valle J, Lopez-Galindo A (2006) Characterization of northern Patagonian Bentonites for pharmaceutical uses. Appl Clay Sci 31:272–281
Viseras C, Aguzzi C, Cerezo P, Lopez-Galindo A (2007) Uses of clay minerals in semisolid health care and therapeutic products. Appl Clay Sci 36:37–50
Viseras C, Cerezo P, Sanchez R, Salcedo I, Aguzzi C (2010) Current challenges in clay minerals for drug deliver. Appl Clay Sci 48:291–295
Wilkinson JB, Moore R (1982) Face packs and masks. In: Wilkinson JB, Moore RJ (eds) Harry's cosmeticology, 7th edn. Longman Group, London
Williams LB, Hillier S (2014) Kaolins and health: from first grade to first aid. Elements 10:207–211
Williams LB, Holland M, Eberl DD, Brunet de Courssou T (2004) Killer clays! Natural antibacterial clay minerals. Mineral Soc Bull 139:3–8
Williams LB, Haydel SE, Giese R, Eberl DD (2008) Chemical and mineralogical characteristics of French green clays used for healing. Clay Clay Miner 56:437–452
Williams LB, Haydel SE, Ferrell R (2009) Bentonite Bandaids and Borborygmi. Elements 5:99–104
Williams LB, Metge D, Eberl DD, Harvey R, Turner A, Prapaipong P, Poret-Peterson A (2011) What makes natural clay antibacterial? Environ Sci Technol 45:3768–3773
Wilson MJ (2003) Clay mineralogical and related characteristics of geophagic materials. J Chem Ecol 29:1525–1547
Young SL (2011) Craving earth. Columbia University Press, New York, 228 pp
Yvon J, Ferrand T (1996) Preparation ex-situ de peloids: Proprietés thermiques, mécaniques et d´échanges. Proc. meeting "Argille Curative", Salice Terme/PV, Gruppo Ital. AIPEA, pp 67–74
Zague V, Almeida Silva D, Baby AR, Kaneko TM, Velasco MVR (2007) Clay facial masks: physicochemical stability at different storage temperatures. J Cosmet Sci 58:45–51

Chapter 9
Provisioning, Recycling, and Trade of Clay

Abstract Natural clay to be used in the field of health and well-being, whether used as it occurs at its natural site, or after extraction from the deposit, is submitted to a refining processing in order to concentrate the finest grain fraction (enriched in clay minerals sometimes associated with organic compounds and finely grained carbonates, potentially the most interesting components in therapeutic or cosmetic terms). In the first situation clay in the form of paste (after being mixed with mineral water, e.g., seawater) is used at its natural site or at a nearby site as mud in mud therapy treatments, or as small pieces chewed and ingested, or even ingested after being dispersed preferentially in potable water providing argillic water. Processed clay is available on the market in two main states: powders (used in pharmaceutical formulations) and paste (used in peloids to be applied, for instance, at spa resorts after undergoing more or less complex manipulation and maturation), and in both states a microbiological check-up is necessary to assure sanitary safety. This sanitary safety is also required when recycling previously used clay paste.

9.1 Purchase, Storage, and Conservation of Clay

Using clays implies that the required material is available both in nature and on the market. In the last situation clay has to be in a state of conditioning chosen by suppliers. The products are supposed to meet certain aims and thus have all the necessary qualities.

Traditionally suppliers of clay get their material in zones close to the surface of the ground where it forms nonhomogeneous reserves. Many villages historically had a small local clay reserve that could have all kinds of uses (such as heat-resistant material for the inside of hearths or fireplaces). These small reserves are not all exhausted and very often others can be found in the vicinity. These clays are obviously not completely suitable because they contain other minerals and impurities; in popular language they are considered "grounds". Nevertheless they can be used in an emergency, which can sometimes be extremely useful if their nature is quickly and precisely determined.[1]

[1] In emergencies, quick recognition of the nature of clay is necessary to avoid making errors. This procedure cannot be approached in this work because it imposes a specific training of the staff and implies long and precise handling.

M. Rautureau et al., *Clays and Health*, DOI 10.1007/978-3-319-42884-0_9

To return to the commercial field, the four principal extracted states suggested are in the form of:

1. Compact pieces: They are often "rough of career". To use them one must crush them or soak them in water. The "ghassoul"[2] is often delivered in this form.
2. Plates: They result from drying after decantation of the purified clay.
3. Powder: It is directly usable, either dry (e.g., in the case of talc) or mixed with water to make a paste. It is the form most often presented by the suppliers who can give it various qualities (fine, very fine, superfine, aired, ventilated, etc.).
4. Paste: The pasty product is directly usable. The water content is variable according to its preparation depending on the type of application required.

The conditions of conservation must take into account the properties of clay and the constraints of the user:

1. *Properties of clay*: Being a mineral ready to hydrate or adsorb many molecules, the place of its storage must preferably be dry and prohibit any contact with toxic solvent molecules or their vapors.
2. *Constraints for users*: For health use, the minimal precaution consists in avoiding any contamination by microorganisms before use. Hydrated clays constitute a sufficient source of water for bacteria, algae, or mushrooms, for example. In principle, in its mass clay is not a receptive medium for these organisms but crushing the clay increases accessibility of microorganisms. Common sense justifies prudence and imposes minimal concepts of hygiene.

There is no essential problem for the conservation of material: it is a stable mineral under normal atmospheric conditions. The essential point to be supervised is that of the variability of the hydration state whose evolution can possibly cause difficulties in its use.

9.2 Initial Preparation by the Supplier

Clay deposits, as a rule, comprise various layers that are texturally, physically, mineralogically, and chemically differentiated by the geological formation environment and by the regional history. In general, the clay layers are sensitive to water; rain brings serious difficulties of extraction and microbiological contamination. The care taken in extraction at the site must make it possible to avoid these potential contaminations:

- The site can be an open-air quarry or an underground one, dry or wet. Clay presents two essential forms, one very compact and almost without porosity and the other rather porous, often mixed with other minerals.

[2] The ghassoul (or rhassoul, rhassoulite) is used for its detergent properties, mainly in North Africa. According to its origin, it is often a stevensite, a little substituted saponite, even a hectorite. These clay minerals belong to the family (or groups) of not very charged trioctahedral smectites; their substitutions are in the octahedral layer and have a soft action because the effect of the electric potentials of these substitution charges is shielded by the two tetrahedral layers.

- The first operation is extraction by explosion, shoveling, powerful water jets, and manual collecting, among others.
- Transport to a storage platform, possibly a second sorting-out and drying.
- The product is then crushed before being put in suspension in water in order to separate the undesirable components: sand, crystallized minerals, stones, no clayey materials, and the like.
- Clay is carried away with the liquid, concentrated, and then dried in order to obtain rather thin plates that can sometimes be sold as they are (ghassoul).
- The purified and dried product can be reworked to give it the desired form (pieces, powder, paste, suspension).

Numerous alternatives of these operations exist. The preparation stage is essential to ensure good quality of the final properties. The microcrystalline character of the clay particles is important and their low dimensions are at the origin of an important part of their properties. These objects are fragile and sensitive to the various preparation phases. In particular, the stages of division of the blocks and setting in suspension are critical and lead to a partial or total amorphization (loss of crystallinity):

- If the mineral is too strongly mechanically crushed
- When crushing is replaced by a cyclone separation (driven by air with very fast rotation) that manages to split up the particles
- When ultrasonic sound, an extremely effective method, causes a progressive amorphization

An over-elaborate division of the quarry clay generally causes a reduction in size and/or number of particle layers and thus increases their proportion of edge layers and consequently of their effects. Their initial properties are disturbed.

It can also appear necessary to the suppliers to carry out a chemical treatment in order to adjust the color, remove residual calcium, remove iron, choose the compensation ions, and so on; the reasons are numerous. None of these operations is neutral for these minerals which are already fragile themselves; they can then lose an important part of their properties.

The drying process is also critical, because it is necessary, from the energy point of view, to remain in a range of temperatures where water losses are reversible. This remark justifies the fact that some producers sun-dry their clays, which works very well and limits the risks of the crystalline structure in particular exceeding the irreversible transformation[3] temperatures (except for halloysite which can remain hydrated only while being maintained in water). However, it is completely superfluous to affirm that clay is able to bring us "the whole of positive energies of the solar rays." These rather current terminologies could nevertheless be studied to know their origins and reasons and, on the other hand, to find which concrete meaning they have for the public at large. For example, an assumption could be that, in addition to a clean drying without the use of energy, solar rays can have a sterilizing effect when the mineral drying particles are not too thick.

[3] These temperatures are higher than 400 °C; in most cases they involve the beginning of the gaseous evolution and elimination of the hydroxyl groups.

Special attention must be brought to the heating process that can seem necessary to purify or sterilize clay. Heating was advanced by certain authorities to cover themselves regarding always possible contaminations. We saw that heat moves water molecules and hydroxyls, thus modifying the crystal structure. Some of these losses, appearing at quite precise temperatures, are irreversible and modify the physicochemical properties completely, leaving the clay without therapeutic interest. The solution is following the product from its removal from quarries down to its use; all these stages must be particularly free from possible contamination. It is the only possibility of preserving the natural aptitudes of these materials.

After these stages of preparation of the product, the packaging should not be neglected. The wrapping intended for the product must have a sufficient quality of impermeability and ensure good insulation to protect the clay from chemical or external microbiological contaminations. It should not be forgotten that clay sooner or later comes in contact with the body in nondeterminable conditions a priori but some of them can definitely be associated with infections or contaminations.

At the final stage of use, sanitary precautions are necessary in the preparation to avoid any contamination by microorganisms: these are common sense precautions.

9.3 Recycling

It is quite difficult to think of recycling clays used in therapeutic applications. The nature of the particles is in question: returning to the initial state of hydration, or chemical element refills, is very delicate and cannot be seriously considered within the framework of domestic use. Moreover, the majority of clays are neither sufficiently rare nor expensive to justify multiple uses.

Let us consider some cases separately.

- *Possible re-using with minor treatment*: When clay was used as cataplasm intended to bring heat to relieve an organ in internal difficulty, it is possible that the same person may use this clay again. In the event of shortage of matter and provided one is sure of the absence of biological, viral, or bacterial contamination, the reuse of this clay can be tolerated. In case of doubt, the mass can be sterilized at temperatures close to 100°C. A higher temperature requires a pressure saturated with steam to maintain water in the paste (use of an autoclave). The temperature of constitutional water loss must not be exceeded because it causes irreversible structural modifications and is at the origin of loss of some properties.

  For the final elimination of material, the return to nature does not pose a problem.
- *In absence of contamination*: There is no contraindication to replace clay in its natural environment. Soil contains a significant amount of clay necessary for the retention of fertilizing elements and a certain quantity of water. Clay scattered around finds its normal place there and the permanent reorganization of the soil will enable it to return to a natural state.

- *After possible or unquestionable contamination*: In this case it is clear that clay material should not be restored to the natural environment nor handled with bare hands. For example, we can mention the medical use of clays such as an antibacterial surface barrier on the skin or to cure serious ulcers.

According to the case, it is possible to eliminate any dangerous trace by strongly heating material that destroys organic traces. The ceramic industry (tiles or bricks) could reuse this clay and recycle it like raw material.

9.4 Marketing

The marketing of clays intended for care must adapt to numerous constraints. It is necessary to be able to:

- Satisfy customers in search of a cure or a physical or psychological well-being and adapt to such sensitivity.
- Find the producer of the raw material.
- Format this raw material with requirements specific to the medical or ancillary medical field.
- Have dynamics of distribution as well as adapted logistics.

All these points do not always have convergent objectives. The use of clays for health still is very directed towards self-medication which introduces few argued choices and, consequently, leaves great importance to tradition. Then the criteria of selection can be distorted or turned away from their objectives. The difficulty is that one does not need "common clay," but one specifically needs a well mineralogically specified clay.

Facing these difficulties it is advisable to give some recommendations.

The origin, nature, and composition of the mineral must in all cases be known to users. It is important to know in great detail how the clay was prepared. It is imperative that the nature of the additives be indicated. Let us recall that certain additives introduced into talc led to very serious and regrettable errors a few decades ago.

In addition, recall that it is advisable to be wary of many unreliable criteria. For example, the color of clay, which is not specific to only one mineral, characterizes the states of ions or impurities that can be associated with various clays.

The transport of clays is no longer a problem with modern technical possibilities. However. some uses, traditional or not, can necessitate looking for autochthonous clays for two reasons:

1. Convenience of local provisioning
2. Existence of medical emergencies, for example, for physicians, local or humanitarian, who must act quickly on the scene of emergencies by giving preference to local possibilities, such as in the event of cholera or in treating drinking water after a natural disaster or war.

9.5 Current Economic Impact

For a long time clay minerals have been elements of the usual pharmacopeia, often in their natural state. Therefore they have been sought after and their value dependent on their possible scarcity and transport difficulties. It is different nowadays. Transport is easy and the quantity of clay used for health is relatively low in regard to transport capacities. The modern pharmacopeia wants to be primarily based on scientific research and it largely uses synthetic techniques of fine chemistry. These techniques are on the whole very expensive and their implication in the socioeconomic balance of medical care is very important in financial terms.

For ideological, economic, ecological, or scientific reasons, "natural products" are again gradually becoming popular after a period of absence, abandonment, or perhaps negligence. Some drug molecules are very delicate to synthesize whereas equivalent natural mineral products are available. In addition, their substitution by natural products in some cases proves extremely effective. However, the use of natural products does not seem to appeal to industrial and financial units associated with the synthesis industry. In addition, natural products are not subject to patents as synthesis products are; thus one can find on the market synthesis products of a level at least equal to natural products but accompanied by much stronger commercial dynamics giving them a prevalent place. In fact, this reasoning is erroneous because when we manage to lower the health budget financially, those efforts on actions that require advanced and complex scientific research are deferred. If this choice is political or financial in favor of the pharmaceutical laboratories, it is each citizen's responsibility. The diffusion of knowledge concerning these products of natural origin remains still strongly marked by their traditional use.

It is necessary nevertheless to be really wary about the name "natural product" which is not universally synonymous with "healthy product",[4] these two notions not being of the same nature.

The knowledge of empirical methods needs to be supplemented by a scientific approach to give them a new statute, which is possible without these materials losing their natural character. The diffusion of this knowledge requires popularizing it and thus a scientific and teaching investment in the users and prescribers, particularly in the field of certain "parallel medicines."

[4] The examples of pollution of the products known as "natural" are quite numerous. The caesium 137 (^{137}Cs) repercussions after the accident at Chernobyl are proof of it. In a more general way, disused mines can contaminate the soil, and thus the clays that they contain, by heavy metals (Pb, As, Ag, etc.). The effluents and rejections of industries reach clays where they may be fixed. These facts must be borne in mind by the public at large and make it possible to judge whether the natural product is wholesome.

Appendix A
Summary of the Relevant Characteristics of Clay Minerals for Therapeutic and Cosmetic Applications

Following is a partial list of the most relevant properties of clay minerals that are of interest in the fields of health and well-being.

They are phyllosilicates of either planar or nonplanar structures.
They are grain size microcrystallines, as a rule, less than 2 micrometers (2 μm).
They have large specific surface areas.
They are ionic compounds.
Three great types of elementary structural layers exist: TeOc, TeOcTe, and TeOcTe-Oc.
The layers often present point defects due to atomic substitutions.
If substitutions create an imbalance of electric charges, compensation is done by interlayer ions.
Generally the interlayer ions are exchangeable: cation exchange capacity and anion exchange capacity exist.
Polar molecules can be fixed on the structural layers or surround and solvate the exchangeable ions.
The fixing and compensation of exchangeable ions or molecules can promote the swelling (volume increase) of clay minerals and, consequently, of clays.
The edges of the structural layers are active zones; they can fix ions or molecules.
The layers are assembled in particles named tactoïdes.
When the layers are not electrically charged or are quasi-neutral, they easily slide over each other (e.g., talc).
Clay minerals are delicately abrasive (tribology).
Clay and clay minerals can form stable gels (thixotropy) and suspensions.
The clay pastes show plasticity and viscosity (rheology).
Clays are sensitive to heat.
The heat storage capacity of a clay paste is high.
Clays and clay minerals can be attacked by strong acids and bases.
Heating at high temperatures can produce irreversible modifications on clay minerals structures

M. Rautureau et al., *Clays and Health*, DOI 10.1007/978-3-319-42884-0

Appendix B
Therapeutic Properties of Clays and Pathologies Treated with Clays

Following is an alphabetically ordered list of clay applications that have been reported both in the popular and scientific press. It is not exhaustive and could be broader regarding publications that could have escaped our reading. In regard to the pathologies treated with clay, a redistribution by classes of application would be essential. Within the framework of this appendix we do not give scientific descriptions allowing a rigorous approach to the use of clay; this description comes from the medical field.

B.1 Properties

• Analgesic	• Cauterization
• Anticarence	• Desinfiltrant
• Antidiarrheic	• Detergent
• Anti-infectious	• Disinfectant
• Anti-inflammatory	• Fixing of odors by adsorption
• Antimycosic	• Great absorptivity and covering
• Antiphlogistic	• Hemolytic
• Antipoison	• Healing
• Antiserotonin	• Hemostatic
• Antitoxin	• Pesticide
• Antedematic	• Regulator of the digestive flora
• Bioactivator (*idem* trace elements)	• Sedative
• Catalytic therapy (*idem* action coppers)	

M. Rautureau et al., *Clays and Health*, DOI 10.1007/978-3-319-42884-0

B.2 Pathologies

• Abscess	• Bulbitis
• Aches	• Burns
• Acne	• Callus
• Aerocolia	• Cancer
• Aerophagia	• Carbuncle
• Ageing	• Cataract
• Alcalosis	• Cellulitis
• Allergy	• Cephalgia
• Alopecia	• Chilblain
• Amenorrhoea	• Cicatrizations
• Amoebiasis	• Cold
• Amoebic dysentery	• Colibacillosis
• Angina pectoris	• Colic
• Angina	• Colitis
• Anorexia	• Comedones
• Arthralgia	• Compress
• Arthritis	• Congestion
• Asthenia	• Conjunctivitis
• Asthma	• Constipation
• Baby	• Contusion
• Bedsore	• Convalescence
• Biliary lithiasis	• Coxarthrose
• Bite	• Cranial pain
• Bleeding nose	• Cutaneous wounds
• Blennorrhoea	• Cyst
• Blepharite	• Cyst, ovarian
• Blood circulation	• Cystitis
• Blood pressure	• Deafness
• Body odors	• Decalcification
• Breastfeeding	• Decay
• Bronchitis	• Demineralization
• Bruise	• Dental abscess, decays

• Depression	• Hemorrhages
• Dermatosis	• Hepatic disorders
• Diabetes	• Hepatitis
• Diarrhea	• Hernia hiatale
• Digestive disorders	• Herpes
• Distensions	• Hydarthrose
• Distorsion	• Hypertension
• Distresses	• Hypocalcemy

(continued)

• Disturbed breath	• Impetigo
• Drainage	• Infantile rednesses
• Drip	• Inflated feet
• Duodenite	• Insect bites
• Dyshidrose	• Insolation
• Dysmenorrhea	• Intestinal worms
• Dyspepsias	• Itching
• Eczema	• Laryngitis (cataplasm gorges)
• Enteritis	• Lipoma
• Epistaxis	• Liver
• Erythema gluteus	• Lumbago
• Evils of belly	• Luxation
• Evils of stomach	• Measles
• Evils of throat	• Medicamentous intoxication
• Eyes	• Melena
• Facial vascular pain	• Menopause
• Fever	• Metrites
• Fibroma	• Migraine
• Fibromyalgia	• Mouth bath
• Flatulence	• Mouth ulcers
• Food poisoning	• Muscular pains
• Fracture	• Myalgia
• Furuncle	• Mycosis
• Gall bladder	• Nails
• Ganglia	• Nephritis
• Gangrene	• Neuralgias
• Gas	• Neuritis
• Gastritis	• Obesity
• Gastroenteritis	• Obstruction of the bowels
• Geophagia	• Ocular ignition
• Skin cracking	• Œsophagites
• Gingivitis	• Ophthalmic migraine
• Glaucoma	• Osteitis
• Glossitis	• Osteoarthritis
• Gums	• Osteochondrosis
• Hematoma	• Osteomalacia
• Hemorrhoids	• Osteoporosis
• Headaches	• Otitis
• Heavy legs	• Overwork
• Hypotension	• Skin cracking

(continued)

• Oxyures	• Spondylarthrite ankylose
• Painful centers	• Stiff neck
• Painful rules	• Stomachic acidity
• Pains of ear	• Stomatitis
• Palpitation	• Stress
• Parasitosis	• Stretch marks
• Parasitosis	• Sty
• Periarticular affection	• Tachycardia
• Peritonitis	• Teeth
• Perleche	• Tendinitis
• Phlegmon	• Thyroid
• Plague	• Tiredness
• Polluted water	• Tonsillitis
• Post operational	• Too strong perspiration
• Pregnancy	• Tooths
• Prurigo	• Toxins
• Pruritus	• Traumatism
• Psoriasis	• Turista
• Purulent buttons	• Ulcerate of Buruli
• Pyoderma	• Ulcerate stomach
• Pyorrhea	• Ulceration
• Rednesses, baby	• Urinary ignition
• Remineralization	• Urticaria
• Removing obstructions from	• Vaccination
• Rheumatisms	• Vaginal discharge
• Rhinopharyngitis	• Vaginal hygiene
• Rhinorrhea	• Vaginitis
• Rickets	• Varicose ulcer
• Scabs	• Varicose vein
• Sciatic nerve	• Various pains
• Scoliosis	• Vitamin deficiency
• Seborrhea	• Warts
• Seize up	• Weakens
• Shingles	• Whitlow
• Sinusitis	• Wounds
• Skin	• Wrench
• Smallpox	• Yearly cracks
• Spasmophilia	

(continued)

B.3 Clay in Well-Being and Beauty

• Alopecia (baldness, loss of hair)	• Body mask
• Bath circulation	• Contour of the eyes (wrinkle, swelling)
• Bath sensitive feet	• Deep wrinkles
• Bath softening dermatosis	• Deodorant or deodorizer
• Bath inflated feet	• Drainage
• Bath painful feet	• Face
• Body care	• Mask greasy skins
• Face pack	• Mask irritated skin
• Gumming body	• Mask normal hair
• Gumming fatty skin	• Mask normal hair
• Gumming normal skin	• Mask normal skin
• Gumming skin dries/dull	• Mask reactive skin
• Hands	• Mask resting
• Invigorating bath	• Mask sensitive skin
• Mask acne	• Mask skin blotches
• Mask capillary	• Mask tonic
• Mask cleansing	• Mask tonic chest
• Mask clarifying	• Nutritive mask
• Mask dry hair	• Releasing bath
• Mask dry skin	• Relieving
• Mask dull skin	• Slimming bath
• Mask fragile hair	• Surface wrinkles
• Mask greasy hair	

Appendix C
Origin of the Names of Clay Minerals

Clay minerals are crystallized natural compounds that do not have true forms and for a long time were without chemical or mineralogical determinations. Nevertheless, their perceptible characteristics distinguished them from other minerals, and sometimes even between them. They were allotted names by prospectors, naturalists, geologists, or mineralogists. With analytical progress, these names very often proved to be redundant, and minerals nomenclature committees, starting from various logical bases, accepted some of them (name of the inventor, remarkable property, place or date of collection, honorary dedication, etc.). However, certain names were kept because they were anchored by their use, but in science only the names adopted by the nomenclature committees were retained.

A recent report by Guggenheim et al. (2006) published in two scientific journals *Clay Minerals* and *Clay and Clay Minerals* shows the names actually attributed to both clay minerals and clays.

In this book that deals with clay and health we consider the various criteria historically used to denominate clay minerals, particularly those interesting for therapeutic and cosmetic purposes, including some names no longer used. References are made to the authors who first discovered and described such clay minerals.

C.1 Criteria Based on the Name of the Mineral Deposit Where the Clay Mineral Occurs

Attapulgite: Mineral described by J. de Lapparent in 1935[1]; this name comes from the locality of Attapulgus (Georgia, United States); it was identified simultaneously in Mormoiron (Vaucluse, France) in the same way with Quincy (Florida, United States); the name attapulgite should be replaced by the name "palygorskite"; the name attapulgite is now used for commercial purposes.

[1] de Lapparent J (1936). Formule et structure de l'attapulgite. *Compt. Rend. Ac. Sc*, Paris, 202:1728–1731.

M. Rautureau et al., *Clays and Health*, DOI 10.1007/978-3-319-42884-0

Beidellite: Name of the Beidell (Colorado, United States) mine given by Larsen in 1925.[2]

Ghassoulite: Grease-removing utility clay used in the Maghreb and exploited in Morocco; intermediary between stevensite and hectorite; exploited in Morocco in Djebel Ghassoul.

Hectorite: Name of the town of Hector (California, United States).[3]

Illite: Name given by Grimm in 1937[4] following a study carried out in Illinois (United States) bearing on "a clay of the mica type."

Kaolinite: The rock bearing "kaolin" resulting from the name of the site, known since 1712, of Kao-Ling in China; the mineral named kaolinite has a more recent name given by Johnson and Blake in 1867.[5]

Montmorillonite: Identified for the first time at Montmorillon (Vienne, France) in 1847 by Damour and Salvetat.[6]

Nontronite: Determined in Nontron (Vienne, France) by Berthier[7] in 1827.

Palygorskite: This clay mineral, of the type "leather of mountain" thus papyraceous, was identified by the Russian Saftchenkov[8] in Palygorsk (Ural mine field, province of Perm, USSR); this name was retained by mineralogists to indicate the attapulgite which has the same crystallochimic formula.

Talc: Name coming from Arabic "talq"; Agricola mentions it in 1546.[9]

C.2 Criteria Based on the Name of a Specific Clay Mineral Property

Chlorite: It represents a family of minerals; name initially given to green mica by Werner in 1799; a family of clay minerals proved thereafter to have similar properties, it was redefined by Pauling[10] (1930) and then by Orcel[11] (1949).

[2] Larsen ES, Wherry ET (1925) Beidellite, has new mineral name. *J. Wash. Acad. Sci.* 15:465–466.

[3] Stress H, Hoffmann U (1941). Synthesis von Magnesiumsilikat-Gelen put zweidimensional regelmäßiger Struktur. *Z. Anorg. Ale. Chem.*, 247:65–95.

[4] Grim RE, Bray RH, Bradley WF (1937) The mica in clay sediments. *Amer. Mineral.*, 22:813–829.

[5] Johnson SW, Blake JM (1867) On kaolinite and pholerite. *Amer. J. Sci.*, 43:351–361.

[6] Damour AA, Salvetat D (1847) Analyse sur un hydrosilicate d'alumine trouvé à Montmorillon. *Ann. Chim. Phys., séries* 3, 21:376–383.

[7] Berthier P (1827) Nontronite, nouveau minéral, découvert dans le département de la Dordogne (France). *Ann. Chim. Phys.*, 36:22–27.

[8] Savchenkov TV (1862) Palygorskit, *Verh. Russ Kaiserl Ges Mineral*, St Petersburg, 102–104.

[9] Agricola, Georgius (2004) De natura fossilium (Textbook of mineralogy). Translated from the first Latin edition of 1546 by Mark Chance Bandy and Jean A. Bandy. Mineola, N.Y.

[10] Pauling L (1930) The structure of chlorites. *Proc. Nat. Acad. Sci., the U.S.A.*, 16:578–582.

[11] Orcel J, Hénin S, Caillère S (1949) Sur les silicates phylliteux des minerais de fer oolithiques. *C.R. Acad. Sci.*, Paris, 229:134–135.

Chrysotile: A very particular mineral with tubular structure and known since antiquity, it was used with the constitution as isolating fabric and refractory materials. Its name, which means "gilded fibers," comes from the Greek *chrysos* = gold and *tilos* = fibers and was given in 1834; it compares to asbestos, the name of a commercial product containing other refractory acicular minerals. It is at the origin of pulmonary mesotheliomas and its use is now prohibited by many countries.

Glauconite: Name coming from the Greek *glaucos* which means green-pale; defined by Brongniart then Humboldt in 1823 and then Keferstein in 1828[12] because of its color; still called glauconite.

Saponite: Described in 1756 by Cronstedt, its name is due to its use as soap; it was studied by Svanberg.[13]

Sepiolite: Known in the past as "sea foam," this light mineral floats on water; the name comes from the Greek *sepion* which means "cuttlefish bone." It was identified by Glocker, 1847.[14]

Smectite: Name with its origin in the Greek term *argos* that means "to clean"; the name (*smectis*) was employed by Cronstedt in 1788. It is currently the name of a family of clay minerals and not that of a well-defined mineral.

C.3 Criteria Based on Homage to Certain Individuals and Entities

Halloysite: Berthier in 1826 gave this name in homage to Omalius d'Halloy to a clay mineral that occurs in the surroundings of Liege (Belgium).[15]
Stevensite: Identified by Leeds in 1873[16] who gave it the name of its institute.

C.4 Critera Based on Technical or Commercial Data

China-Clay: Kaolin employed by the ceramics industry.
Floridine: Fuller's Earth, bentonite described by Spencer in 1925.
Fuller's Earth: Very varied localizations, often a mixture of smectites and palygorskite; used to degrease wool and the skin.
Ground of Sommières: Extracted from quarries located at Salinelles (Gard, France); mixture of sepiolite, saponite, and hectorite; used as a stain-remover.

---

[12] Keferstein C (1828) Deustch Geognotisch. Geol. dargestellt., 5,3, p 510.

[13] Svanberg LF (1840) Saponit och Rosit, tvenne nya mineralier. K Vetensk Akad Stockholm Handl 153–166 & *Ann. Phys. Chem.* 2. F. 57, 165–170.

[14] Werner G (1789) Letztes Mineral System—Herausgegeben von C Hoffmann *Bergmännisches J.*, 1:377.

[15] Berthier P (1826) Analyse de l'halloysite. *Ann. Chim. Phys.* 32:332–335.

[16] Leeds AR (1873) With hydrous unisilicate approaching pyrosclerite *Amer. J. Sci.*, 6:22–24.

Appendix D
Clays and Clay Minerals Used in the Field of Health

This appendix presents a list of clay minerals of practical interest in the field of health, a list that is not exhaustive and can be supplemented by the readers. There is an essential difficulty of geological and mineralogical nature. The concepts of rock and clay mineral must be differentiated. Sometimes a clayey rock can be made up of only one clay mineral or of one major clay mineral, the rest being some other trace minerals, but very often a clayey rock consists of a mixture of various clay minerals associated with other types of minerals. This leads to distinguishing, for example, kaolin (rock) from kaolinite (clay mineral). The attapulgite of Mormoiron (France), for example, is a mixture of palygorskite and montmorillonite.

The distinctions between clayey rocks and clay minerals are clearly fields of the specialists (geologists, mineralogists, and crystallographers) to whom the various actors involved in the commercial uses must refer. Clarity and information remain an obligation. An example of this insufficient characterization can be found in the notes associated with a drug prescription, such as "typical clay mineral molecule" similar to montmorillonite, the distinction between the molecular state and the crystallized state being completely forgotten. The fact represents a double problem, that of language, and that of formation. Undoubtedly the problem will gradually be solved.

D.1 Natural Clay Minerals Used in the Field of Health

Names considered clay minerals (alphabetical order):

- Beidellite
- Chlorites (family of chlorites)
- Hectorite
- Illite
- Interstratified clay minerals
- Kaolinite
- Montmorillonite

M. Rautureau et al., *Clays and Health*, DOI 10.1007/978-3-319-42884-0

- Nontronite
- Palygorskite (including attapulgite)
- Saponite
- Sauconite
- Sepiolite
- Stevensite
- Talc

Names not corresponding to clay minerals, but to rocks containing clay minerals:

- Bentonite
- Ghassoul, rhassoul
- Kaolin
- Ochre (goethite, hematite, and kaolinite)

Some examples of common names (often corresponding to specific places or to specific uses):

- Asbestos
- Fuller's Earth
- Ground or earth of Carpentras (France)
- Ground or earth of Sommières (France)

Denominations without much interest:

- Ionized clay
- Red, green, white clay, and so on
- Heavy clay
- Activated clay

D.2 Pharmaceutical Commercial Products Containing Clay Minerals

The following are only some of the names of commercial pharmaceutical products, sometimes old or already disregarded; there are very many other names.

It is advisable to distinguish the names attributed to drugs from those names attributed to daily care products employed, for example, in beauty care. Clay and clay minerals are used either as active ingredients or as excipients (Carretero 2009). Similarly the veterinary use of clay and clay minerals could be taken into account.

- Actapulgite®
- Actrys® *bandages*
- Aloplastine®
- Argel 7®
- Arligea®

- Bedelix®
- Eryplast®
- Escargil®
- Gastropax®
- Gastropulgite®
- Gelox®
- Illite
- Kaobrol®
- Kaolin
- Kaologeais®
- Kaopectate®
- Kapanol®
- Karayal®
- Montmorillonite
- Mucipulgite®
- Mulkine®
- Octalite®
- Palygorskite
- Clay water paste
- Smecta®
- Stomargil®
- Talc

Appendix E
The Surface Area of Clays

Although possibly appearing complex to certain readers, it seemed necessary for us to introduce a structured approach of the background pertinent to the knowledge of the extremely important properties of clays, the surface area, and the specific surface area.

We often discussed how important the surface area of a clay is. It is a physical parameter we must treat in detail; this would be only to justify, by a simple calculation, the announced values but also the many properties and applications of health associated with the surface of clays. The values below are fundamental to understanding the role of clays' surface areas in their therapeutic actions.

Let us suppose that the particles are cubic and the edge is *a*.

Their volume is a^3 and their surface $6a^2$.

For *n* particles, the volume of the sample is $V_e = na^3$ and its surface $= 6na^2$.

Specific surface known per unit of volume: $S_u = \frac{S_e}{V_e} = \frac{6na}{na^3} = \frac{6}{a}$.

If ds is the specific gravity of the material, the surface per unit of weight is: $S_u = \frac{S_e}{V_e \mathrm{ds}} = \frac{6}{a \cdot \mathrm{ds}}$.

For example, with $a = 0.01$ cm, in a volume of 1 cm^3, the total surface of the particles is:

$S_e = 600\ \mathrm{cm}^2$

For a density equal to 2.5 the result is that we have a ponderal specific surface of 240 $\mathrm{cm}^2\ \mathrm{g}^{-1}$.

By imagining the spherical particles, the same result would be obtained, the value of the diameter *D* replacing that of the edge of the cube.

One of the essential properties of phyllite minerals is its formation by plates: thus let us consider a right prism thickness *Z* and sides *X* and *Y*. By applying the preceding method of calculating, the formula becomes: $S_u = \frac{S_e}{V_e} = \frac{2(xy + yz + zx)}{xyz} = 2\left(\frac{1}{x} + \frac{1}{y} + \frac{1}{z}\right)$.

M. Rautureau et al., *Clays and Health*, DOI 10.1007/978-3-319-42884-0

For clays, one knows that x and y are large there, compared to z, and have a negligible contribution to the value of the specific surface. The order of magnitude of this surface is thus fixed by the value of $S_u \approx 2/z$.

Consequently, if we calculate for a cube of 1 cm on side, formed of equidistant planes of z along one of its edges: the number of sections is given by $1/z$ and each section (a layer) reveals two surfaces: the entire surface is thus practically equal to $2/z$. For example, for a thickness of 1 nm (i.e., 10^{-7} cm), $S_u = 2.10^7$ cm^2 = 2000 m^2. For a density of 2.5 one obtains a specific surface of 800 m^{2}g^{-1}. This result is very close to reality in the case of montmorillonite.

If an identical operation is carried out on an edge perpendicular to that used previously, the volume is divided into very lengthened prisms, or sticks, and its surface is doubled: $S_u = \frac{S_e}{V_e} = \frac{2}{x} + \frac{2}{y} + \frac{2}{z} \approx \frac{4}{z}$.

However, the number of the sticks is now equal to 10^{14}.

By reproducing this operation on the x-axis the contact surface is tripled and the number of particles becomes $(1/z)^3 = 10^{21}$.

These remarks take all their value if the growth of the crystals is studied; a reduction of 10^{21} to 10^7 (i.e., 10^{14} times) of the number of the particles and for layers of 1 nm thickness, corresponds only to one 66% reduction of the surface.

These values are checked practically by measuring the quantity of fluid adsorbed under standard conditions. Gas nitrogen can be used with the BET method (developed by S. Brunauer, P. H. Emmett, and F. Teller, and described in a paper entitled "Adsorption of Gases in Multi-Molecular Layers" published in *J. Amer. Chem. Soc.* (1938), 60:309–319), a method that gives external surfaces, or with the glycol method (by Hendricks), which gives the total contact surface, internal and external, in the case of smectites whose layers can be separated.

Appendix F
Small Dictionary of Clay

The word "clay" has a principal translation in all languages, and many equivalent terms as well that refer to folk and traditional uses and properties. The readers are asked to supplement it.

English:

- Clay
- Potter's earth
- Clay ground or clay soil
- Loam
- Mud
- Suspension

French:

Clay	Argile
Clay ground	Terre
Loam	Terre glaise
Mud	Boue
Suspension	Suspension

Latin:

Clay	*Argilla*
Clay	*Lutum*, which also indicates mud and potter's earth (the word gave lute (coated hardening) and the verb to lute (a pot!)
Ground	
Loam	
Mud	In addition to *lutum* the word *caenum* also exists
Suspension	*Aqua lutea*, that is, muddy water (*luteus* being the adjective derived from *lutum*)

M. Rautureau et al., *Clays and Health*, DOI 10.1007/978-3-319-42884-0

Greek:

Clay	*Argillos* (or *argilos*); cf. the adjective *argilôdès*: clayey)
Clay	
Ground	
Loam	*Kéramos* (potter's earth; but also any terra cotta object; cf. the "Keraméion" = district of the potters in Athens)
Mud	*Pèlos* (exactly: slurry; but sometimes designates loam; cf. the adjective *pèlodès* = muddy)
Suspension	

Arabic:

Clay	طين	*tîn*
Clay (hard)	صلصال	*salsâl*
Ground	أرض	*'ard*
Loam		
Mud Mud (malleable)	وحل حما	*wahl* *hamâ*
Suspension	علق البحر	*'alaq al-bahr*

Persian:

Clay	خاک رس	*khâk-e ros*
Clay	گل رس	*gel-e ros*
Ground	خاک	*khâk*
Loam		
Mud	گل	*gel*
Suspension		

Spanish:

Clay	Arcilla
Clay	Barro, greda
Ground	Tierra
Loam	Greda
Mud	Barro, lodo
Suspension	Suspensión

Portuguese:

Clay	Argila
Clay	Barro, greda
Ground	Terra
Loam	Barro
Mud	Lama, lodo, limo
Suspension	Suspensão

Appendix G
Original International Classification of pelods, in French, Established in Dax, France, in 1949

The International Classification of Peloids (the original, in French, is shown in Table G.1) was established and also approved at the Fourth Conference of the ISMH (International Society of Medical Hydrology) that was held in Dax, France, in 1949. More than 65 years later it is still the official classification being based on the following factors: origin of the peloid solid component, chemical nature and temperature of mineral water, and maturation conditions of the mixture.

Table G.1 Classification internationale originale des Péloïdes (Dax, 1949)

Dénomination des Péloïdes	Origine	Eau minéral		Conditions de Maturation
		Nature chimique	Température	
Boues (Fanghi, Muds, Schlamm)	A prédominance inorganique (minérale)	Sulfureuse, sulfatée, chlorurée, bromurée, iodurée	Hyperthermale *au griffon*	(a) In situ (sur le *griffon* des sources)
			Homéothermale (36–38°C) *au griffon*	(b) En bassin
			Hypothermale *au griffon*	
Limons	Id.	Eau de mer ou de lac salé	Hyperthermale *au bassin*	In situ
Tourbes (Torbe, Peats, Moor)	A prédominance organique	Alcaline, carbonatée, ferrugineuse, sulfureuse	Hyperthermale *au griffon*	(a) Ouverte
		Eau de mer	Homéothermale *au griffon*	(b) Couverte
			Hypothermale *au griffon*	
			Hypothermale *au bassin*	
Muffes (Mousses, Barégines)	Id.	Sulfureuse	Hyperthermale *au griffon*	in situ

(continued)

M. Rautureau et al., *Clays and Health*, DOI 10.1007/978-3-319-42884-0

Dénomination des Péloïdes	Origine	Eau minéral		Conditions de Maturation
		Nature chimique	Température	
Bioglées autres que le muffe (Algues, etc.)	Id.	Eaux minérales autres que les eaux sulfureuses	Hyperthermale *au griffon*	Id.
			Homéothermale *au griffon*	
			Hypothermale *au griffon*	
Sapropeli	Mixte	Alcaline, ferrugineuse, sulfureuse	Hypothermale *au bassin*	Id.
Gyttja	Id.	Eau de mer	Id.	Id.

Appendix H
Simplified Transliteration Adopted in Arabic and Persian

We have adopted a simplified system similar to that found in certain recent publications (Cheddadi, Abdesselam, *Arabs and the Appropriation of the History*, Sindbad, Actes Sud, Paris, 2004). We have transliterated the two letters *hamza* and *ayn* while using for both the right apostrophe. We do not mark the signs indicating the emphatic ones and the velar ones. Thus, Arabic-speaking people will be able to find the conventional transcription, and non-Arabic-speaking people will have easier reading.

M. Rautureau et al., *Clays and Health*, DOI 10.1007/978-3-319-42884-0

Table H.1 Codes of the simplified transliteration for Arabic and Persian

Arabic letters	Simplified transliteration in this work	International transliteration
أ	’	’
ا	â	ā
ث	th	ṯ
ج	dj	ǧ
ح	h	ḥ
خ	kh	ḫ
ذ	dh	ḏ
ش	sh	š
ص	s	ṣ
ض	d	ḍ
ط	t	ṭ
ظ	z	ẓ
ع	‘	‘
غ	gh	ġ
ق	q	q
و	û- w	ū
ي	î– y	ī
Additional letters for Persian		
پ	p	-
چ	tch	-
ژ	j	-
گ	g	-
و	u-v-o	-
ِ - ا	e	-

Important remark:

In this book, sometimes we indicate the word “clay” in English and sometimes in Arabic language where the term “ طين ” means clay.

Using the correct international transliteration, this term should be written as “*ṭīn*”, the letter “t” should have a point below. However, to simplify the lecture for **non-Arabic-speaking** readers, we used simplified transliteration “*tîn*”, by using a normal “t” of Latin alphabet, without a point below.

Thus, we have to give a precision for **Arabic-speaking readers** for whom this simplification may seem a mistake for the rough pronunciation, wrong spelling and wrong meaning. In this book, when sometimes we use the Arabic word “*tîn*”, it refers to the translation “clay,” not the translation “fig” although the simplified transliteration is the same for both meanings. To avoid the criticisms of the Arabic-speaking readers and to reassure them, the table below gives the necessary explanation.

Arabic writing	**Simplified Transliteration in this book**	**Correct International Transliteration**	Significance
طين	*tîn* In this book, we used this term to indicate the clay, in Arabic language.	*ṭīn* The correct transliteration of the Arabic written form requires the point under the letter "t". However, in this book, we do not use it and we always use simplified transliteration.	clay
تين	*tîn*	*tīn*	fig In this book, we are not concerned about this fruit!

Appendix I
Photographic and Graphic Credit

Figure 4.4	Nicole Liewig, IPHC/DEPE, Strasbourg, France
Figures 5.6b and 5.7b	Alain Baronnet, Paul Cézanne University & Centre Interdisciplinaire de Nanoscience de Marseille-CNRS, UPR 3118, Campus Luminy, Marseille, France Équipe traceurs minéraux du Centre de Recherche en Matière Condensée et Nanosciences
Figure 5.10	Nadia Alem: Sonatrach/Centre de Recherche et de développement avenue du 1er novembre, Boumerdès 35000, Algérie
Figure 5.11a–c	Olivier Dequincey (thesis, 2001) & Nicole Liewig, CGS, Strasbourg
Figure 7.1 Figure 8.9	François Reichardt (thesis, 2008) & Nicole Liewig: IPHC, Strasbourg
Figure 6.6a	1990. Anne-Marie Jaunet: INRA, Versailles, France
Figure 6.6b	2008. Toshihiro Kogure, Department of Earth and Planetary Science, The University of Tokyo
Figure 6.7	Hervé Gaboriau (Thesis, Université d'Orléans), BRGM, Orléans, France
Figures 4.1, 4.2, 4.3, 5.1, 5.2, 5.8, 5.9, 5.13, 5.14, 6.1, 6.2, 6.3, 6.4, 6.5, 8.1, 8.8	Michel Rautureau, author
Dessins 5.3, 5.4, 5.5, 5.6a, 5.7a, 5.12	Michel Rautureau with permission of Société des Industries Minérales, images extracted from «Les argiles» (2004)

M. Rautureau et al., *Clays and Health*, DOI 10.1007/978-3-319-42884-0

Glossary

Abrasion Wear due to extraction of a fraction of matter by a slicing effect resulting from a difference in hardness and an adapted geometry.

Absorption Absorption by clay occurs inside the space located between the layers of the particles (interlayer space); penetration of a substance in an organ by passage through the membranes of the organ (digestive tract, skin, mucous membrane) without causing lesion.

Adsorption Phenomenon of fixation of gas or liquid molecules on solid surfaces; it should not be confused with absorption; in the case of a clay the adsorption is done on the external surface of the particles and their edges.

Amorphous, amorphization State of a material that in part or as a whole does not exhibit the ordered tridimensional organization proper to the crystalline state.

Calorie Whose symbol **cal**, is a unit of energy, defined (calorie in 15°C) as being the quantity of heat (or energy) necessary to raise the temperature of 1 g of pure water of 1°C (of 14.5°C with 15.5°C) to the normal atmospheric pressure (1 atm is 101.325 kPa); the calorie unit should no longer be used, but it persists because its mental representation is simple; the international system imposed its replacement by the joule (symbol: J) with conversion 1 cal (15°C) = 4.18 J.

Clay A naturally occurring geomaterial composed primarily of fine-grained minerals —the so-called *clay minerals* being the fundamental constituents—that is generally plastic at appropriate water content and hardens on drying or firing.

Clay ground or clay soil Ground or soil impermeable to water able to exhibit higher or less plasticity.

Clay milk or "argillic clay" The suspension surviving after decantation of a mixture of clay in a strong quantity of water; it can be used as a drug.

Clay minerals Hydrous phyllosilicates, fundamental constituents of clays.

Colloid Mix or dispersion of an aqueous solution in a liquid solvent, the particles having a size larger than that of the molecules of the solvent.

Colloidal Dispersion or mixture in a liquid solvent of an aqueous solution made up of particles whose size is larger than that of the molecules of the liquid; colloidal dispersions have the property to diffuse light.

M. Rautureau et al., *Clays and Health*, DOI 10.1007/978-3-319-42884-0

Crystal, crystalline State of the solid matter that results from the rigorous geometrical arrangement of the constituent atoms of a material.

Crystalline structure Describes the tridimensional ordered organization of the atoms of a crystal.

Decked (clay) Clay modified by intercalation of pillars between the layers.

Decrystallization, decrystallized Loss of the tridimensional organized character of the atoms of a crystalline solid.

Dehydration Loss of the water molecules associated with interlayer or interlayer solvated or hydrated ions.

Dehydroxylation Loss or elimination of hydroxyls from the octahedral layers through water vapor evolution when a clay is heated to high temperature; there is a collapse or destruction of clay minerals structures before the formation of the so-called high-temperature phases.

Dialysis Method of separation of a component contained in a liquid by osmosis through a membrane.

Drying Process carried out at low temperature to eliminate interstitial free water, below 100°C until obtaining a constant weight; for higher temperatures, the occurrence of the phenomenon called dehydration, or even the phenomenon called dehydroxylation must be taken into account.

Edible clay Clay that is deliberately ingested by man as cookies or as clay/potable water dispersions; it can act as a source of mineral micronutrients and as protective material against pathogens and toxins (detoxification of noxious or unpalatable compounds present in the diet), or it can act as an antacid for gastric acidity compensation and neutralization.

Essential oil Concentrated and hydrophobic liquid, of the evaporative aromatic compounds of a plant, obtained by distillation.

Extemporaneous peloid Healing mud that is prepared locally at the time of use, blending one common or one special clay (kaolin or bentonite), or algae with natural mineral water or seawater without undergoing maturation; it is not a real peloid *s.s.* (*stricto sensu*) because maturation does not occur.

Fat clay This term often appears to describe a sufficiently humidified and plastic clay.

Filter candle Filters consisting of a porous ceramic whose pore dimension limits that of the filtered products; often used in the field when it is impossible to employ average sophisticated techniques; they can be associated with activated carbon or exchanging resins with ions.

Gel State of a mixture or suspension that lost its liquid properties because of internal physicochemical interactions.

Geophagia Deliberate soil-eating, earth-eating, clay-eating, and pica; it has been observed in all parts of the world since antiquity, reflecting cultural practice, religious belief, and physiological needs, whether nutritional (dietary supplementation) or as a remedy for disease. Goes back to prehistoric times, and is widespread across the world.

Ground or soil (in pedology) Actual surface part of terrestrial surface resulting from the deterioration of the bedrock and organic material decomposition; the grounds support the cultures well; a current order of magnitude of their thickness is close to 1 meter.

Gumming Action of moving an object out by fixing it by adherence on a mobile support.

Healing or curative clay Clay characterized by healing properties and used by humans for internal or oral applications in the form of cookies or clay/water dispersion; in the form of *mud* clay could also be used for external or topical applications (*mud therapy* and *peloid therapy*); *edible clay* is a particular type of *healing clay* used only for oral applications.

Hydration Water reintroduction in a structure previously drained by physical or chemical action.

Hydrogen potential, pH Measures the acidity of a solution; represents the decimal co-logarithm concentration in hydrogen ions: $\mathbf{pH = -log\,[H^+]}$.

Hydroxyl Noted OH^-; a group or pair of oxygen and hydrogen atoms carrying a negative charge, participating in the constitution of some molecules and crystals.

Interstratified The structural layer type is representative of a clay mineral; to form a particle of such mineral several layers, all identical, have to be piled up by a natural crystal growth process; in nature commonly the layers that form one particle are not all identical (e.g., a clay mineral particle can be constituted of layers of kaolinite regularly or irregularly intercalated with layers of smectite); when this happens this clay mineral is called interstratified or mixed layered in order to differentiate it from the clay mineral constituted of layers of only one type.

Ion An atom or group of atoms electrically charged (by excess or deficit of one or several electrons).

Joule Legal unit of energy that replaces the calorie (see above).

Loam Type of sediment or soil composed of sand, silt, and clay whose properties are intermediate between the extremes of the three components.

Mud The state of a natural water/clay mixture that may contain organic matter, and that can become deformed and move spontaneously under its own weight; it can be used in mud therapy.

Mud Therapy The topical use of natural muddy sediment under the form of mud paste for therapeutic and/or cosmetic purposes, mostly on an empiric way and without medical counselling and surveillance; such natural muddy sediment that has been maturated in its geologic site of occurrence is simply called natural mud (or just mud) or natural peloid.

Osmosis Action that allows the passage between two media separated by a membrane of an atom, a molecule, or an ion whose concentrations are different on both sides of the membrane.

Parapeloid or paramud or parafango Healing peloid or healing mud or healing fango that is the product of the blend of a clay (e.g., bentonite or kaolin) with paraffin; parapeloid is not a type of peloid, otherwise it is a peloid succedaneum.

Pascal-second Dynamic viscosity of a fluid when inside it a linear and uniform movement along a plane surface, solid and infinite, generates a force of friction of $1\ N{\cdot}m^{-2}$ when the movement has a speed of $1\ m{\cdot}s^{-1}$.

Paste State of a water–clay mixture that becomes deformed due only to the action of one force.

Peloid Natural product (natural sediment) composed of a mixture of mineral water (seawater and lake water included) with organic or inorganic matter that results from either geologic or biologic processes or yet from both processes, and is utilized for therapeutic purposes in the form of packs, compresses, or baths; peloids could be prepared for medical use (therapeutic peloid) or for cosmetic use (cosmetic peloid).

Physiological serum Isotonic solution of sterile water; the concentration of 9% NaCl in the physiological serum is identical to that of the liquids of the human body.

Plant oil Produced by pressure or crushing of a crop product of seeds, pips, or fruits.

Plasticity In a paste of water and clay, plasticity results from the nature of the clay particles, made up of layers, that slide over each other.

pH tampon Solute that controls the pH of a medium.

Poise Unit of viscosity.

Poultice or cataplasm Clay paste topically applied onto the skin of certain regions of the body, such as knee, hand, and lower back to treat rheumatic pathologies.

Proton Noted H^+.

Rehydration Return of water molecules to the crystalline structure of clay minerals; that implies a preliminary dehydration and the existence of a reversible mechanism.

Rheological properties Flow property of a water–clay mixture.

Rheology Refers to the properties of flow of a paste, attributed to the phenomena of viscosity, plasticity, and elasticity.

Seawater Water containing approximately 35 $g \cdot L^{-1}$ of sodium chloride on average; comparatively the physiological serum contains 9 $g \cdot L^{-1}$ of sodium chloride.

Suspension A liquid containing the solid particles that do not undergo the effect of gravity and do not elutriate; they remain suspended in the liquid.

Swelling The insertion of certain molecules between the structural layers of a clay mineral can increase the interlayer distance; consequently the volume of material containing the clay mineral will increase and the measurement of the volume variation provides important information on the mechanism and the origin of the swelling that is primarily due, depending on its uses in the field of health, to the hydration of interlayer cations; many other causes exist, however, to explain the swelling.

Texture The visible shape of the components of a material, clay in layer, fibrous clay.

Thermal conductivity The aptitude to be transferred from heat through a material.

Thermal mud Result after maturation of a mud that can contain organic matter.

Thixotropy Property of forming a gel; property of the viscous fluids that become liquid when agitated or when a mechanical constraint is applied to them, and again become viscous after a certain rest period.

Trace elements Chemical elements (simple) in the organism in extremely reduced quantity but essential to its life; the trace elements are metals or metalloids with a precise role in the biological processes; they are essential to protein activity and can intervene as enzymatic cofactors; they can, in addition, intervene as hormonal cofactors or help to stabilize the structure of nucleotides or nucleic acids (DNA); their absence can be the cause of health disorders; if present in too high content they can be toxic.

Tyxotropic properties Aptitude of a water–clay mixture to form a gel.

Viscosity Property that characterizes the capacity of an incompressible fluid to flow (the poise presently replaced by Pascal-second (Pa·s) is the unit that measures viscosity).

Further Reading

Aznar A et al (1992) Colloid Polym Sci 270:165–176

Bergaya F, Theng BKG, Lagaly G (eds) (2013) Handbook of clay science. Elsevier, New York

Carretero MI, Pozo M (2009) Clay and non-clay minerals in the pharmaceutical and cosmetic industry, Part I Excipients and medical applications. Appl Clay Sci 46:73–80; Part II Active ingredients, Applied Clay Science, 47:171–181

Collectif, coordination Estrade J (1994) Argiles. Laboratoires Beaufour, Paris

Dequincey O (2001) Apport des déséquilibres 238u-234u-230th à l'étude minéralogique et géochimique des profils d'altération, cas de deux profils latéritiques africains, thèse, Strasbourg 1

Favier A (1991) Les oligoéléments en nutrition humaine. In: Chappuis P (ed) Les oligoéléments en médecine et biologie. Editions Médicales Internationales, Lavoisier, Paris

Favier A, Lamond M (1999) Les oligoéléments, éd. Tec. & Doc, CNRS-CNERNA, pp 41–49

Foucault A, Raoult JF (2005) Dictionnaire de géologie. 6ème édition, Dunod, Paris

Gaboriau H (1991) Interstratifiés smectite-kaolinite de l'Eure: relations entre la structure, la texture et les propriétés en fonderie. Thèse, Sciences Physiques Appliquées, Université d'Orléans, Orléans

Gilardi JD, Duffy SS, Munn CA, Tell LA (1999) Biochemical functions of geophagia in parrots: detoxification of dietary toxins and cytoprotective effects. J Chem Ecol 25:897–922

Ibn al-Baytâr (1992) Traité des Simples, traduction de Leclerc L, éd. Institut du monde arabe, Paris, 3 volumes

Laszlo P (1990) Les argiles en chimie organique. La Recherche 219(21):314–323

Lozet J, Mathieu C (2002) Dictionnaire des sciences du sol, Lavoisier éd. Tech & Doc, 4ème éd.

Meunier A (2003) Argiles, éditions scientifiques G.B, Paris

Newmann ACD (1987) Chemistry of clays and clay minerals, vol 6. Mineralogical Society, London

Pomerol C, Ricour J (1992) Terroirs et thermalisme de France, ed. BRGM

Reichardt F (2008) Ingestion d'argile spontanée chez le rat: rôle dans la physiologie intestinale, Thèse, Université Louis Pasteur, Strasbourg

Robertson RHS (1961) Mineral use guide, ed. Cleaver-Hume Press, London

Van Damme H, Lemaire E, Levitz P (1991) La physique de la boue. Pour la Science 167:82–90

Velde B (1995) Origin and mineralogy of clays. Springer, Berlin

Yvon J, Thomas F, Villiéras F, Michot LJ (1993) Surface activity of water. In: Julia D et al (eds) Power technology of pharmaceutical process. Handbook of power technology, vol 9. Elsevier, Amsterdam, pp 79–113

Reading of General Interest, in French

Actes Sud, collectif (2009) Argiles—Histoire d'avenir, éd. Actes Sud

Aedis éditions (2007) L'argile et ses bienfaits
Alma-Helal A (1998) Guérir par l'argile, éditions Godefroy
Andrianne P (2003) L'argile, médecine ancestrale, éd. Amyris-Inspir coll. Médecine douce, Bruxelles
Beauvais M (2001) Magie de l'argile, éd. La maison rustique
Binet C (1991) Oligoéléments et oligothérapie, éditions Dangles
Bourgeois P (1981) Soignez-vous avec l'argile, éditions De Vicchi, Paris
Carité JM, collectif, (2008) L'argile médicinale pour votre santé et votre beauté, éditions Utovie
Clergeaud C (1999) L'argile, terre de vie, éditions Labussière, Tournus, France
Cousin N (2013) Argile, édition Eyrolles
Coutaret G (2003) La guérison par l'argile, éditions Bussière, Paris
Deixtreit R (1997) L'argile qui guérit, éditions Vivre en Harmonie
Derlon P (1978) La médecine secrète des gens du voyage, Robert Laffont
Donnadieu Y (1980) L'argile: thérapeutique naturelle, éditions Maloine (collection: les thérapeutiques naturelles)
Guenot JP (1984) Comprendre l'argile, Editions EFIC
Guenot JP (1989) Les argiles: comment choisir, Editions Chiron
Guillain F (2006) L'argile, tout simplement, éditions Demeter
Huard L (2004) L'argile au service du corps, Quebecor (collection santé naturelle)
Kotchenko N (1989) Les 7 couleurs de l'argile soleil. Editions Argiletz
Kotchenko N (2003) Les bienfaits de l'argile. Editions Marabout
L'homme et l'argile, http://lhomme.et.largile.free.fr/
Mantovani R (1975) L'argile terre miraculeuse. Editions Mantovani
Masson R (1986) La santé par l'argile, éditions Solar
Muller MF (1998) L'argile facile: une thérapie naturelle millénaire. Editions Jouvence
Nigelle E (1996) Pouvoirs merveilleux de l'argile. Editions Andrillon
Olivier JF (1988) L'argilothérapie: les argiles et leurs compléments naturels, éd. Encre
Passebecq A (2007) L'argile pour votre santé. Editions Dangles
Plante L (2005) Comment se soigner par l'argile, éditions Edimag
Skavinska V (1985) Le livre de l'argile. Editions Michel Laffont

Some Scientific Societies Dedicated to Clays

Association Internationale pour l'Étude des Argiles (AIPEA): http://aipea.org/
Clay Minerals Society (CMS): cms@clays.org
European Clay Groups Association: http://www.ing.pan.pl/ecga_js/ecga1.htm#site_map
Groupe Français des Argiles: http://www.gfa.asso.fr/

Index

M. Rautureau et al., *Clays and Health*, DOI 10.1007/978-3-319-42884-0

MIX
Papier aus verantwortungsvollen Quellen
Paper from responsible sources
FSC® C105338

If you have any concerns about our products,
you can contact us on
ProductSafety@springernature.com

In case Publisher is established outside the EU,
the EU authorized representative is:
Springer Nature Customer Service Center GmbH
Europaplatz 3, 69115 Heidelberg, Germany

Printed by Libri Plureos GmbH
in Hamburg, Germany